Edmund B. Weston

Report of the Results Obtained with Experimental Filters

At the Pettaconset Pumping Station of the Providence Water Works

Edmund B. Weston

Report of the Results Obtained with Experimental Filters
At the Pettaconset Pumping Station of the Providence Water Works

ISBN/EAN: 9783337139087

Printed in Europe, USA, Canada, Australia, Japan

Cover: Foto ©berggeist007 / pixelio.de

More available books at **www.hansebooks.com**

APPENDIX TO THE SEVENTEENTH ANNUAL REPORT OF THE
STATE BOARD OF HEALTH OF RHODE ISLAND,
FOR THE YEAR ENDING DEC. 31, 1894.

REPORT

OF THE

RESULTS OBTAINED

WITH

EXPERIMENTAL FILTERS

AT THE PETTACONSET PUMPING STATION OF THE PROVIDENCE WATER WORKS.

BY EDMUND B. WESTON,

MEMBER OF THE AMERICAN SOCIETY OF CIVIL ENGINEERS.
MEMBER OF THE INSTITUTION OF CIVIL ENGINEERS OF GREAT BRITAIN.
ASSISTANT ENGINEER IN CHARGE OF WATER DEPARTMENT, PROVIDENCE, R. I.

PROVIDENCE, R. I.:
E. L. FREEMAN & SONS, STATE PRINTERS.
1896.

GENERAL CONTENTS.

PAGE.

REPORT OF THE RESULTS OBTAINED WITH EXPERIMENTAL FILTERS AT THE PETTACONSET PUMPING STATION OF THE PROVIDENCE WATER WORKS.

By EDMUND B. WESTON, C.E.,
Assistant Engineer in Charge of Water Department.

City Engineer's Office, Water Department.
Providence, R. I., March 12, 1894.

Mr. J. Herbert Shedd, City Engineer.

Dear Sir:—As directed by you, I commenced the experimental filtration work in the Cornish Engine House at Pettaconset Pumping Station in February, 1893.

The experimental filters, having been set up, were started about March 27, and the work was continued, with the exception of a brief interval in September, until January 30, 1894.

Two Experimental Filters, Nos. 1 and 2, built according to your design, and an Experimental Morison Mechanical Filter, were used during the course of the work. Another Experimental Mechanical Filter was also set up, at the expense of the owners of the same and run for several months, but as the results obtained with this filter were not satisfactory, it will not again be mentioned in this report.

Nos. 1 and 2 filters were run for about seven months as natural and mechanical filters, at rates of flow of about 2,000,000, 5,000,000 and 30,000,000 gallons per acre per 24 hours. The work with these filters was then discontinued.

The term Natural Filtration, is used in this report to designate the filtration of water through a bed of sand or quartz, when there has not been any foreign substance added to the Applied Water, and the term Mechanical Filtration, is used to designate the filtration of water through a bed of sand or quartz, when some foreign substance, such as Basic Sulphate of Alumina, has been added to the Applied Water.

A sketch representing Filters Nos. 1 and 2, is shown in Cut No. 1.

The arrangement of the interior of each of the Filters Nos. 1 and 2, is as follows :—

Upon the bottom of the filter a bed of cement about six (6) inches deep was laid. Several rows of bricks, set on edge, upon the cement, about six (6) inches apart supported a floor of bricks laid flatwise. The bricks of the floor were cut so as not to leave any openings of more than one-sixteenth ($\frac{1}{16}$) of an inch wide in the joints between the bricks or at the sides of the filter. The filter-bed consisted of three (3) inches of "pea gravel," which was placed upon the brick floor, a layer of coarse sand one (1) inch thick, which was laid upon the top of the "pea gravel," and upon this coarse sand the filtering medium proper was placed, which was composed of a layer of "fine sand," of uniform quality, one (1) foot and eight (8) inches deep. During the experiments several kinds of "fine sand" were used at different times in Filter No. 1 as the filtering medium and one kind during the entire time in Filter No. 2, the Effective Size and Uniformity Coefficient of which, are given in table No. 1.

A representation of the Experimental Morison Mechanical Filter, which is described in detail in that portion of the report which describes the experiments made with this filter, is shown in Cut No. 2.

At the end of the runs of each of the Filters, Nos. 1 and 2, the filter-bed was either washed, or about one-half ($\frac{1}{2}$) of an inch of the top of the filtering medium scraped off.

When each filter-bed was washed a one-inch hose was connected to the discharge-pipe at *A*. This hose supplied the water used in washing, which was forced up through the brick floor and filter-bed under pressure, and overflowed through the six-inch branch at *B*, the cap, to which the inlet-pipe *C* is connected, having been removed.

The results that were obtained with Filters Nos. 1 and 2, and the chemicals that were used when they were run as mechanical filters are given in table No. 1. In the same table are given the results that were obtained with the Experimental Morison Mechanical Filter during the time that Filters Nos. 1 and 2 were in service. This table is intended for comparison only, as the bacterial colonies, from which the percentages contained in the table were determined, were only cultivated from 42 to 62 hours, and "Fifteen-

Cut No. 1.

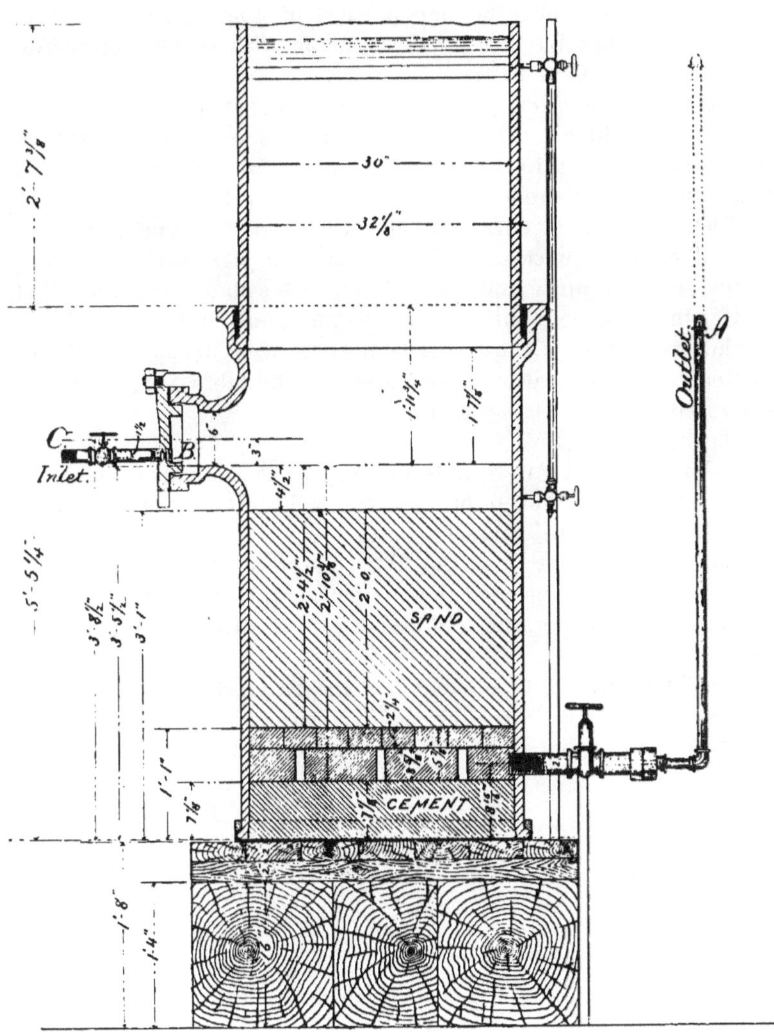

Experimental Filters.

Nos. 1 and 2

per-cent Gelatin" was used the greater part of the time as the cultivating medium instead of "Ten-per-cent Gelatin," consequently, it is quite likely that the percentages given in the table are slightly unreliable. The methods followed in regard to the bacteriological work are described in detail under the head of Bacteriological Work in that portion of the report which describes the experiments that were made with the Experimental Morison Mechanical Filter.

Table No. 1 practically describes itself, and it contains the only information which I shall hereafter mention in regard to Filters Nos. 1 and 2, owing to the very limited time at my disposal for the preparation of this report.

When Alumina or Alumina and "Free Flow," are mentioned in the table, in connection with the filters, it signifies that *Basic Sulphate of Alumina* and "Free Flow," were added to the Applied Water in the same manner as will be mentioned later on in detail in the description of the Experimental Morison Mechanical Filter. When Natural Filtration is mentioned in the table it signifies that the filters were run as natural filters.

The "Effective Size" and the "Uniformity Coefficient" mentioned in the table, were determined by the methods followed at the Experiment Station of the Massachusetts State Board of Health at Lawrence, Massachusetts.

The "Effective Size" of the filter material (diameter in millimeters). This size is such that 10 per cent. by weight of the material is of smaller grains, and 90 per cent. is of larger grains than the size given.

The "Uniformity Coefficient" is the ratio A to B when the values of A and B are such that 60 per cent. by weight of the material is finer than A and 10 per cent. finer than B.

TABLE No. 1.

Showing a comparison of the Results obtained from Filters Nos. 1 and 2 and the Morison Mechanical Filter from March 27 to October 5, 1893. The Samples of Filtered Water in the Table were all taken at the Same Hour as the Samples of Applied Water. The Bacteria were cultivated from 42 to 62 Hours.

DATE.	Number of Bacteria in Applied Water.	FILTER No. 1.			FILTER No. 2.			MORISON MECHANICAL FILTER.		
		Number of Bacteria in Filtered Water.	Per cent. of Applied Water Bacteria Removed.	Gallons of Water Filtered per Acre, per 24 Hours.	Number of Bacteria in Filtered Water.	Per cent. of Applied Water Bacteria Removed.	Gallons of Water Filtered per Acre, per 24 Hours.	Number of Bacteria in Filtered Water.	Per cent. of Applied Water Bacteria Removed.	Gallons of Water Filtered per Acre, per 24 Hours.
		Effective Size of the Sand Grains of the Filtering Medium 0.81 mm. Uniformity coefficient 2.2. Started Filters Nos. 1 and 2. Natural Filtration.								
March 27, 1893..	....	2,795	....	3,110,000	196	....	2,100,000			
28, " ..	....	7,930	....	2,830,000	672	....	1,640,000	The average length of a run of the Morison Mechanical Filter (between washings), during the period covered by this table was about 18 hours. River Water was used for washing the Filter Bed.		
29, " ..	....	533	....	4,910,000	278	....	8,400,000			
30, "	....	650	...	4,730,000	3,610	....	860,000			
31, " ..	683	481	29.6	4,390,000	422	38.2	570,000			
April 1, " ..	858	334	61.1	4,390,000	500	41.7	5,700,000			
2, " ..	...	....	....	....	....	....				
3, " .	569	103	81.9	4,670,000	585	....	1,230,000			

TABLE NO. 1.—CONTINUED.

DATE.	Number of Bacteria in Applied Water.	FILTER No. 1.			FILTER No. 2.			MORISON MECHANICAL FILTER.		
		Number of Bacteria in Filtered Water.	Per cent. of Applied Water Bacteria Removed.	Gallons of Water Filtered per Acre, per 24 Hours.	Number of Bacteria in Filtered Water.	Per cent. of Applied Water Bacteria Removed.	Gallons of Water Filtered per Acre, per 24 Hours.	Number of Bacteria in Filtered Water.	Per cent. of Applied Water Bacteria Removed.	Gallons of Water Filtered per Acre, per 24 Hours.
April 4, 1893...	614	464	24.4	1,600,000	152	75.2	1,580,000	Effective Size of the Quartz Grains of the Filtering Medium 0.59 mm. Uniformity coefficient 1.5.		
								Started Filter. Natural Filtration.		
5, " ...	789	222	71.9	1,540,000	419	46.9	1,220,000	329	58.3	....
6, " ...	808	123	84.8	1,450,000	52	93.6	1,830,000	169	79.1	128,000,000
7, " ...	587	2,145	....	1,360,000	56	90.5	2,080,000	....	...	...
8, " ...	315	234	25.7	1,280,000	97	69.2	1,160,000	13	95.9	96,000,000
9, " ...	....	....	...	....	....	....	....	....	...	...
10, " ...	2,030	222	89.1	1,930,000	440	78.3	1,810,000	2,395	....	118,000,000
11, " ...	....	....	....	....	...	....	...	....		....
12, " ...	983	113	88.5	1,370,000	79	92.0	1,250,000	744	24.3	127,000,000
13, " ...	....	....	....	....	....	....	....	....		....
									Alumina.	
14, " ...	923	64	93.1	1,700,000	74	92.0	1,560,000	520	43.7	61,500,000

April 15, 1893..	1,527	20	98.7	1,820,000	11	99.3	1,470,000	7	99.5	110,000,000
16, " ..	...	....	....		....	....	....	...	....	
17, " ..	1,508	13	99.1	1,740,000	12	99.2	1,520,000	15	99.0	127,000,000
18, " ..	971	21	97.8	2,120,000	6	99.4	1,560,000	229	76.4	132,000,000
19, " ..	962	27	97.2	1,250,000	27	97.2	1,250,000	281	70.8	178,000,000
20, " ..	1,439	....	...	....	19	98.7	2,120,000	377	73.8	127,000,000
21, " ..	3,055	....	...	....	16	99.5	2,160,000	6	99.8	73,000,000
		Filter Bed Repacked with New Sand. Effective Size of Sand Grains 0.18 mm.								
22, " ..	3,965	....	....	....	88	97.8	3,730,000	390	90.2	127,000,000
		Uniformity coefficient 2.1.								
23, " ..	...		....	...	....	....	....	...	....	....
		Started Filter. Natural Filtration.								
24, " ..	892	....	....	....	10	98.9	1,740,000	390	56.3	131,000,000
25, " .	966	170	82.4	4,920,000	10	99.0	1,740,000	119	87.7	115,000,000
								Alumina and "Free Flow."		
26, " ..	1,228	....	....	....	11	99.1	2,080,000	1	99.9	116,000,000
27, " ..	2,138	366	82.9	4,750,000	12	99.4	2,430,000	11	99.5	118,000,000
28, " ..	2,144	103	95.2	3,380,000	13	99.4	1,600,000	14	99.3	110,000,000
29, " ..	2,268	..	...	....	15	99.3	2,900,000	12	99.5	123,000,000

Table No. 1.—Continued.

Date.	Number of Bacteria in Applied Water.	Filter No. 1.			Filter No. 2.			Morison Mechanical Filter.		
		Number of Bacteria in Filtered Water.	Per cent. of Applied Water Bacteria Removed.	Gallons of Water Filtered per Acre, per 24 Hours.	Number of Bacteria in Filtered Water.	Per cent. of Applied Water Bacteria Removed.	Gallons of Water Filtered per Acre, per 24 Hours.	Number of Bacteria in Filtered Water.	Per cent. of Applied Water Bacteria Removed.	Gallons of Water Filtered per Acre, per 24 Hours.
		Washed Bed with Filtered Water. Started Filter. Natural Filtration.								
April 30, 1893 ...	....	....		....	....		....	....	....	....
May 1, " ...	1,907	106	94.4	2,020,000	42	97.8	1,470,000	29	98.5	116,000,000
2, " ...	1,417	125	91.2	1,820,000	10	99.3	1,200,000	5	99.6	121,000,000
3, " ...	1,988	24	98.8	1,740,000	12	99.4	1,660,000	4	99.8	119,000,000
4, " ...	2,310	20	99.1	1,620,000	32	98.6	2,360,000	3	99.9	93,000,000
5, " ...	7,831	...	....	....	77	99.0	1,520,000	18	99.8	106,000,000
6, "	1,634	....		....		....		5	99.7	113,000,000
7, " ...	....		....	....		....			....	....
8, "	707					....	....	5	99.3	106,000,000
9, " ...	767	....		....	....	....	....	1	99.9	103,000,000

Date										
May 10, 1893...	906	....	....	....	....	....	....	6	99.3	108,000,000
		Filter Bed Repacked with New Sand. Effective Size of Sand Grains 0.35 mm.								
11, " ..	....	....	....	....	...			....	....	....
		Uniformity coefficient 2.0.			Filter Bed Repacked with Same Sand.			Natural Filtration.		
12, " ...	1,569	...	....	....	....	....	...	308	80.4	128,000,000
		Started Filter.		Natural Filtration.	Started Filter.		Natural Filtration.			
13, " ...	819	2,684	...	2,360,000	3,767	...	1,190,000	453	44.7	171,000,000
14, " ...	....	....	....	...	....		....	....	....	....
								Alumina and "Free Flow."		
15, " ...	1,352	2,456	...	2,050,000	598	55.8	2,120,000	2	99.9	152,000,000
16, " ...	442	4,134	....	2,900,000	639		3,650,000	...	....	....
17, " ...	3,662	814	77.8	2,400,000	519	85.8	2,160,000	...	....	....
18, " ...	2,649	221	91.7	1,990,000	188	92.9	2,090,000	....	....	....
19, " ...	1,946	165	91.5	2,950,000	2,524	....	1,400,000	....	....	....
20, " ...	1,551	253	83.7	2,560,000	448	71.1	2,030,000	....	....	....
21, " ...	....	....	....	....	....	....	....	....	....	....
22, " ...	334	240	28.1	3,350,000	192	42.5	2,320,000	....		....
23, " ...	2,112	58	97.3	1,880,000	185	91.2	2,630,000	....	....	....
24, " ...	4,054	1,425	64.8	1,740,000	150	96.3	2,020,000	...	....	....

Table No. 1.—Continued.

Date.	Number of Bacteria in Applied Water.	Filter No. 1.			Filter No. 2.			Morison Mechanical Filter.		
		Number of Bacteria in Filtered Water.	Per cent. of Applied Water Bacteria Removed.	Gallons of Water Filtered per Acre, per 24 Hours.	Number of Bacteria in Filtered Water.	Per cent. of Applied Water Bacteria Removed.	Gallons of Water Filtered per Acre, per 24 Hours.	Number of Bacteria in Filtered Water.	Per cent. of Applied Water Bacteria Removed.	Gallons of Water Filtered per Acre, per 24 Hours.
May 25, 1893...	3,090	55	98.2	1,420,000	130	95.8	1,520,000		....	....
26, " ...	2,918	24	99.2	2,500,000	46	98.4	2,240,000	....	....	....
27, " ...	5,135	26	99.5	1,610,000	31	99.4	1,720,000	....	....	....
28, " ...	....	...	....	....	..	....	....	....	...	....
29, " ...	1,858	77	95.9	3,520,000	137	92.6	3,520,000	...	....	....
30, " ...	3,119	55	98.2	2,060,000	34	98.9	2,090,000	...	....	....
31, " ...	9,945	195	98.0	3,220,000	372	96.3	2,000,000	...	....	....
									Alumina.	
June 1, " ...	6,825	381	94.4	1,800,000	323	95.3	2,370,000	3	100	134,000,000
2, " ...	8,202	62	99.2	1,870,000	94	98.9	3,030,000	2	100	131,000,000
3, " ...	18,817	133	99.3	3,020,000	159	99.2	2,843,000	104	99.4	132,000,000
4, " ...	....	....	....	....	...	...	....	...	....	....

Date										
June 5, 1893...	8,445	283	96.6	3,400,000	95	98.9	2,310,000	....		....
6, " ...	10,595	182	98.3	2,430,000	160	98.5	2,170,000	43	99.6	141,000,000
7, " ...	14,950	68	99.5	1,950,000	57	99.6	1,680,000	....	....	....
8, " ...	18,655	44	99.8	1,920,000	125	99.3	2,090,000	....	....	....
9, " ...	8,060	27	99.7	2,200,000	30	99.6	2,320,000	....	....	....
10, " ...	....	....	....	....	....	..	....	....	....	....
11, " ...	....	....	....	....	....	....	....	....	....	....
								Alumina and "Free Flow."		
12, " ...	5,460	8	99.9	1,950,000	14	99.7	2,180,000	0	100	123,000,000
13, " ...	12,187	282	97.7	1,940,000	16	99.9	2,980,000	74	99.4	148,000,000
14, " ...	54,400	....	....	....	4,420	91.9	4,550,000	3	100	125,000,000
		Started Filter.	Scraped Bed.	Natural Filtration.						
15, " ...	66,982	10,452	84.4	3,300,000	1,205	98.2	2,670,000	....	....	....
									Alumina.	
16, " ...	50,825	237	99.5	3,413,000	58	100	2,700,000	2,786	94.5	119,000,000
								Alumina and "Free Flow."		
17, " ...	43,095	591	98.6	3,460,000	146	99.7	2,780,000	1,355	96.9	121,500,000
18, " ...	....	....	....	....	....	....	....	....	....	....
19, " ...	15,015	744	95.0	1,700,000	28	99.8	2,310,000	75	99.5	117,000,000

Table No. 1.—Continued.

Date.	Number of Bacteria in Applied Water.	Filter No. 1.			Filter No. 2.			Morison Mechanical Filter.		
		Number of Bacteria in Filtered Water.	Per cent. of Applied Water Bacteria Removed.	Gallons of Water Filtered per Acre, per 24 Hours.	Number of Bacteria in Filtered Water.	Per cent. of Applied Water Bacteria Removed.	Gallons of Water Filtered per Acre, per 24 Hours.	Number of Bacteria in Filtered Water.	Per cent. of Applied Water Bacteria Removed.	Gallons of Water Filtered per Acre, per 24 Hours.
June 20, 1893...	16,700	180	98.9	3,540,000	54	99.7	2,230,000	63	99.6	136,000,000
21, " ...	40,787	134	99.7	2,230,000	73	99.8	2,070,000		....	....
22, " ...	12,707	19	99.9	2,330,000	36	99.7	1,790,000	...		....
23, " ...	41,855	11	100	2,400,000	12	100	1,920,000			....
24, " ...	20,522	13	99.9	2,200,000	16	99.9	2,030,000	....	...	....
25, " ...	....	....	....	...	....	....	...	....	...	....
26, " ...	1,488	29	98.1	2,850,000	26	98.3	3,210,000	....	....	....
27, " ...	2,762	9	99.7	2,610,000	105	96.2	2,570,000	181	93.4	131,000,000
28, " ...	5,460	11	99.8	2,430,000	190	96.5	2,320,000	8	99.9	123,000,000
29, " ...	3,900	10	99.7	2,060,000	206	94.7	1,750,000	1	100	120,000,000
30, " ...	2,957	2	99.9	2,020,000	17	99.4	1,870,000	7	99.8	132,000,000

July 1, 1893...	4,745	57	98.8	3,020,000	76	98.4	2,330,000	1	100	120,000,000
2, " ...	....	....		....	....	....	....	...	...	....
3, " ...	2,717	112	95.9	1,500,000	150	94.5	2,060,000	1	100	116,000,000
4, " ...	....	....		....	....	....	...	....	....	....
5, " ...	2,567	49	98.1	1,230,000	24	99.1	1,980,000	0	100	128,000,000
								Alumina.		
6, " ...	506	13	97.4	2,720,000	54	89.3	1,740,000	60	88.1	156,000,000
								Alumina and "Free Flow."		
7, " ...	1,319	35	97.3	2,160,000	33	97.5	2,010,000	0	100	136,000,000
8, " ...	944	18	98.1	1,300,000	9	99.0	2,000,000	1	99.9	135,000,000
9, " ...	...	....	....		....	....	...	....	....	....
10, " ...	315	18	94.3	1,490,000	30	90.5	4,820,000	....	....	....
11, " ..	655	4	99.4	1,910,000	15	97.7	2,270,000	1	99.8	125,000,000
								Natural Filtration.		
12, " ...	864	4	99.5	3,300,000	28	96.8	2,250,000	127	85.3	24,000,000
13, " ...	832	27	96.8	1,620,000	40	95.2	1,430,000	519	37.6	23,900,000
14, " ...	599	11	98.2	1,030,000	....	....	....	185	69.1	27,300,000
					Washed Bed with Filtered Water. Started Filter. Natural Filtration.					
15, " ...	1,043	10	99.0	1,490,000	985	5.6	30,200,000	656	37.1	30,200,000

Table No. 1.—Continued.

Date.	Number of Bacteria in Applied Water.	Filter No. 1. Number of Bacteria in Filtered Water.	Filter No. 1. Per cent. of Applied Water Bacteria Removed.	Filter No. 1. Gallons of Water Filtered per Acre, per 24 Hours.	Filter No. 2. Number of Bacteria in Filtered Water.	Filter No. 2. Per cent. of Applied Water Bacteria Removed.	Filter No. 2. Gallons of Water Filtered per Acre, per 24 Hours.	Morison Mechanical Filter. Number of Bacteria in Filtered Water.	Morison Mechanical Filter. Per cent. of Applied Water Bacteria Removed.	Morison Mechanical Filter. Gallons of Water Filtered per Acre, per 24 Hours.
July 16, 1893...	....	....	....	....	....		....	....	....	....
		Scraped Bed. Started Filter. Natural Filtration.								
17, " ...	565	153	72.9	29,300,000	40	92.9	30,500,000	27	95.2	21,300,000
								Alumina and "Free Flow."		
18, " ...	4,030	2,210	45.2	19,700,000	3	99.9	27,300,000	2	100	116,000,000
19, " ...	7,247	780	89.2	28,300,000	27	99.6	28,300,000	2	100	96,000,000
20, " ...	9,067	32	99.6	28,300,000	10	99.9	29,500,000	4	100	125,000,000
21, " ...	7,182	281	96.1	30,500,000	67	99.1	29,300,000	767	89.3	122,000,000
22, " ...	6,890	373	94.6	33,200,000	53	99.2	29,500,000	147	97.9	113,000,000
23, " ...	....	....	....	....	....	....	....	....	....	....
24, " ...	435	5	98.9	29,700,000	4	99.1	28,900,000	1	99.8	125,000,000
					Washed Bed with Filtered Water. Started Filter. Alumina and "Free Flow."					
25, " ...	4,160	....	....	....	49	98.8	27,300,000	1	100	134,000,000

		Scraped Bed Twice. Started Filter. Alumina.								
July 26, 1893..	17,485	2	100	20,700,000	633	96.4	29,400,000	1	100	122,000,000
		Scraped Bed. Started Filter. Alumina.								
27, " ..	15,340	565	96.3	30,600,000	186	98.8	26,500,000	4	100	123,000,000
		Scraped Bed. Started Filter. Alumina and "Free Flow."			Washed Bed with Filtered Water. Started Filter. Alumina and "Free Flow."					
28, " ..	8,542	35	99.6	17,800,000	....	...	....	9	99.9	30,500,000
		Scraped Bed. Started Filter. Alumina and "Free Flow."								
29, " ..	6,417	29	99.5	17,800,000	1,137	82.3	27,300,000	65	99.0	31,500,000
		Filter Bed Repacked with New Sand. Effective Size of Sand Grains 0.35 mm.			Alumina Discontinued. Natural Filtration.					
30, " ..	...	....	...	....	....	....	....		....	...
		Uniformity coefficient 2.0.								
31, " ..	747	....	...	..	60	92.0	26,500,000	19	97.5	24,000,000
		Started Filter. Natural Filtration.								
August 1, " ..	3,900	9,977	....	26,400,000	49	98.7	28,300,000	1	100	128,000,000
2, " ..	11,272	15,942	....	30,500,000	116	99.0	29,500,000	48	99.6	136,000,000
3, " ..	3,282	237	92.8	28,400,000	18	99.5	30,800,000	1	100	112,000,000
		Washed Bed with River Water. Started Filter. Alumina and "Free Flow."								
4, " ..	7,475	73	99.0	26,300,000	107	98.6	33,300,000	14	99.8	121,000,000
5, " .	5,557	78	98.6	29,000,000	162	97.1	29,400,000	4	99.9	132,000,000
		Washed Bed with River Water. Started Filter. Natural Filtration.								
6, "	....	....	....	....	....	....		...	....	...

TABLE NO. 1.—CONTINUED.

DATE.	Number of Bacteria in Applied Water.	FILTER No. 1.			FILTER No. 2.			MORISON MECHANICAL FILTER.		
		Number of Bacteria in Filtered Water.	Per cent. of Applied Water Bacteria Removed.	Gallons of Water Filtered per Acre, per 24 Hours.	Number of Bacteria in Filtered Water.	Per cent. of Applied Water Bacteria Removed.	Gallons of Water Filtered per Acre, per 24 Hours.	Number of Bacteria in Filtered Water.	Per cent. of Applied Water Bacteria Removed.	Gallons of Water Filtered per Acre, per 24 Hours.
August 7, 1893..	461	60	87.0	24,000,000	174	62.3	25,500,000	4	99.1	125,000,000
8, " ..	1,397	107	92.3	31,900,000	232	83.4	29,400,000	4	99.7	125,000,000
9, " ..	7,085	85	98.8	31,900,000	43	99.4	28,300,000	0	100	116,000,000
		Washed Bed with River Water. Started Filter. Natural Filtration.								
10, " ..	16,900	1,950	88.5	30,500,000	134	99.2	28,300,000	0	100	121,000,000
11, " ..	3,315	207	93.8	28,900,000	182	94.5	29,500,000	5	99.8	125,000,000
12, " ..	4,582	167	96.4	26,500,000	224	95.1	34,500,000	28	99.4	125,000,000
13, " ..	...	....	....	....	....	....	...	....	....	....
14, " ..	395	331	16.2	33,300,000	68	82.8	28,300,000	4	99.0	128,000,000
		Filter Bed Repacked with New Sand. Effective Size of Sand Grains 0.81 mm.								
15, " ..	910	....	....	....	50	94.5	28,300,000	21	97.7	121,500,000
		Uniformity coefficient 2.2. Started Filter. Alumina and "Free Flow."								
16, " ..	356	137	61.5	31,900,000	18	94.9	28,300,000	5	98.6	121,500,000

August 17, 1893.	792	64	91.9	31,900,000	13	98.4	29,000,000	9	98.9	121,000,000
18, "	492	27	94.5	28,300,000	32	93.5	33,200,000	0	100	121,500,000
19, "	156	14	91.0	30,600,000	31	80.1	30,600,000	2	98.7	121,500,000
		Washed Bed with River Water. Started Filter. Alumina and "Free Flow."								
20, "	....	....	....	....	....			....	....	....
21, "	523	154	70.6	28,300,000	45	91.4	30,600,000	9	98.3	29,400,000
22, "	1,199	120	90.0	30,800,000	29	97.6	31,200,000	46	96.2	29,800,000
23, "	4,387	278	93.7	31,900,000	....	....		11	99.7	28,300,000
					Scraped Bed. Started Filter. Natural Filtration.					
24, "	....	64	....	28,900,000	....	....	....	1	....	28,000,000
		Washed Bed with River Water. Started Filter. Alumina and "Free Flow."								
25, "	....	....	....	....	253		13,800,000	4	....	28,350,000
26, "	852	59	93.1	28,300,000	81	90.5	30,700,000	114	86.6	29,000,000
27, "	..	....	...	..	...	...		...	....	....
		Washed Bed with River Water. Started Filter. Alumina and "Free Flow."								
28, "	457	109	76.1	27,500,000	56	87.7	29,500,000	...	....	
					Scraped Bed. Started Filter. Natural Filtration.					
29, "	1,939	337	82.6	30,500,000	523	73.0	21,200,000	....	....	....

TABLE No. 1.—CONTINUED.

DATE.	Number of Bacteria in Applied Water.	FILTER No. 1.			FILTER No. 2.			MORISON MECHANICAL FILTER.		
		Number of Bacteria in Filtered Water.	Per cent. of Applied Water Bacteria Removed.	Gallons of Water Filtered per Acre, per 24 Hours.	Number of Bacteria in Filtered Water.	Per cent. of Applied Water Bacteria Removed.	Gallons of Water Filtered per Acre, per 24 Hours.	Number of Bacteria in Filtered Water.	Per cent. of Applied Water Bacteria Removed.	Gallons of Water Filtered per Acre, per 24 Hours.
					Scraped Bed. Started Filter. Natural Filtration.					
Aug. 30, 1893...	1,502	678	54.9	27,500,000	581	61.3	31,900,000	....	....	....
		Washed Bed with River Water. Started Filter. Alumina and "Free Flow."								
31, " ...	6,208	30	99.5	31,800,000	2,145	65.4	33,000,000	....	....	....
Sept. 1, " ...	495	65	86.9	29,500,000	80	83.8	31,800,000	....	....	....
		Washed Bed with River Water. Started Filter. Alumina and "Free Flow."								
2, " ...	1,007	566	43.8	30,600,000	54	94.6	34,500,000	....	....	....
3, " ...	....	....	....	....	..	....	...	....	....	
		Washed Bed with River Water. Started Filter. Alumina and "Free Flow."			Washed Bed with River Water. Started Filter. Natural Filtration.					
4, " ...	576	79	86.3	27,500,000	63	89.1	30,500,000	....	....	....
5, " ...	788	40	94.9	29,500,000	78	90.1	34,500,000	2	99.7	154,000,000
		Washed Bed with River Water. Started Filter. Alumina and "Free Flow."								
6, " ...	1,192	81	93.2	30,500,000	79	93.4	31,800,000	4	99.7	125,000,000

Sept. 7, 1893...	683	444	35.0	31,800,000	14	98.0	29,800,000	1	99.9	122,000,000
8, " ...	1,126	299	73.4	30,500,000	92	91.8	31,800,000	7	99.4	125,000,000
		Washed Bed with River Water. Started Filter. Alumina and "Free Flow."								
9, " ...	4,325	200	95.4	25,500,000	68	98.4	30,500,000	2	100	122,000,000
10, " ...	....	....	....	....	....	....	....	....		
		Washed Bed with River Water. Started Filter. Alumina and "Free Flow."								
11, " ...	470	25	94.7	33,000,000	9	98.1	30,500,000	3	99.4	128,000,000
12, " ...	741	10	98.7	33,000,000	20	97.3	31,800,000	....		....
		Washed Bed with River Water. Started Filter. Alumina and "Free Flow."								
13, " ...	2,467	325	86.8	31,800,000	16	99.4	33,000,000			....
		Washed Bed with River Water. Started Filter. Alumina and "Free Flow."			Washed Bed with River Water. Shut down Filter.					
14, " ...	16,825	2,047	87.8	30,000,000	....	....	....	....	....	....
								Shut down Filter.		
15, " ...	1,040	345	66.8	31,800,000	....	....	....	...		...
		Washed Bed with River Water. Shut down Filter.								
16, " ...	....	....	...	....	....	....	....	....	....	...
17, " ...	....	....	....	....		....		....	....	....
18, " ...	....									...

Table No. 1.—Continued.

Date.	Number of Bacteria in Applied Water.	Filter No. 1. Number of Bacteria in Filtered Water.	Filter No. 1. Per cent. of Applied Water Bacteria Removed.	Filter No. 1. Gallons of Water Filtered per Acre, per 24 Hours.	Filter No. 2. Number of Bacteria in Filtered Water.	Filter No. 2. Per cent. of Applied Water Bacteria Removed.	Filter No. 2. Gallons of Water Filtered per Acre, per 24 Hours.	Morison Mechanical Filter. Number of Bacteria in Filtered Water.	Morison Mechanical Filter. Per cent. of Applied Water Bacteria Removed.	Morison Mechanical Filter. Gallons of Water Filtered per Acre, per 24 Hours.
Sept. 19, 1893...	....	..					...	...		
20, " ...	....		....			....			....	....
21, " ...			....							....
22, " ...	....	....	....							....
23, " ...	....			...	....		....	....	...	....
24, " ...	....	....	....	...			...	...		....
25, " ...	....	....	....	....		....			....	....
26, " ...	....	....	....	....	....	....			....	
		Washed Bed with River Water. Started Filter. Alumina and "Free Flow."			Washed Bed with River Water. Started Filter. Natural Filtration.			Started Filter. Alumina and "Free Flow."		
27, " ...	10,492	5,557	47.0	34,500,000	10,887	....	31,800,000	2	100	126,000,000
		Washed Bed with River Water. Started Filter. Alumina and "Free Flow."								
28, " ...	8,872	1,235	86.1	33,100,000	4,775	46.2	31,800,000	5	99.9	125,000,000

		Washed Bed with River Water. Started Filter. Alumina and "Free Flow."								
Sept. 29, " ...	56,587	18,330	67.6	31,800,000	18,005	68.2	31,800,000	17	100	128,000,000
30, " ...	64,960	17,135	73.6	31,800,000	1,217	98.1	31,800,000	6	100	125,000,000
Oct. 1, " ...	...	....	....	....		....		....	....	....
		Washed Bed with River Water. Started Filter. Alumina and "Free Flow."								
2, " ...	6,857	3,152	54.0	31,800,000	103	98.5	30,500,000	29	99.6	133,000,000
3, " ...	3,683	119	96.8	33,100,000	36	99.0	31,800,000	4	99.9	125,000,000
		Washed Bed with River Water. Started Filter. Alumina and "Free Flow."								
4, " ...	2,632	5,070	....	31,800,000	71	97.3	31,800,000	7	99.7	128,000,000
		Washed Bed with River Water. Started Filter. Alumina and "Free Flow."			Washed Bed with River Water. Started Filter. Natural Filtration.					
5, " ...	1,430	49	96.6	30,000,000	159	88.9	33,200,000	3	99.8	131,000,000
		Shut down Filter.			Shut down Filter.					

The number of bacteria given in the different columns of the table is the number found in one cubic centimeter of water.

TABLE No. 1.—CONTINUED.

Loss of Head in feet, due to the Flow of the Water through the Filter Bed and Discharge Pipe, and the Equivalent Height that the water rose in the filter during each run, after the filter first commenced to discharge at its full capacity, etc.

Date of commencement of run.	FILTER No. 1.				
	Height of discharge-pipe in feet, above the filter-bed, at the commencement of the run.	Loss of Head in feet, due to the flow of the water through the filter-bed and discharge-pipe, at the commencement of each run.	The Equivalent Height in feet that the water rose in the filter during each run, after the filter commenced to discharge at its full capacity.	Gallons of Water Filtered per Acre, per 24 Hours, when the filter commenced to discharge at its full capacity, at the time that the loss of head was measured; also, the average flow of the filter during each run.*	Actual length of each run in days, hours, and minutes.
	Discharge-pipe stationary. The water gradually rose in the filter as the bed clogged up. Effective Size of the Sand Grains 0.81 mm. Uniformity coefficient 2.2. Natural Filtration.				
March 27, 1893	0.00	0.0156	0.0244	2,000,000 (2,276,000)	23 *d.* 4 *h.* 40 *m.*
	Filter Bed Repacked with New Sand. Effective Size of the Sand Grains 0.18 mm. Uniformity coefficient 2.1. Natural Filtration.				
April 24, "	0.00	0.3646	1.5754	5,000,000 (5,764,000)	4 *d.* 6 *h.* 14 *m.*
	Washed Bed with Filtered Water. Natural Filtration.				
April 29, "	0.00	0.1667	1.6633	2,000,000 (1,543,000)	5 *d.* 5 *h.* 10 *m.*
	Water kept in filter at a constant height of about 4.0 feet above the bed. The discharge-pipe was gradually lowered as the bed clogged up. Filter Bed Repacked with New Sand. Effective Size of the Sand Grains 0.35 mm. Uniformity coefficient 2.0. Natural Filtration.				

May 13, 1893..........	4.00	0.06	6.04	2,000,000 (1,700,000)	31 *d.* 5 *h.* 58 *m.*
		Scraped Bed. Natural Filtration.			
June 15, "	4.00	0.02	6.08	2,000,000 (1,930,000)	30 *d.* 7 *h.* 46 *m.*
		Scraped Bed. Natural Filtration.			
July 17, "	4.00	0.72	4.84	30,000,000 (29,200,000)	6 *d.* 22 *h.* 33 *m.*
		Scraped Bed Twice. Alumina.			
July 26, "	4.00	2.30	2.15	30,000,000 (31,200,000)	0 *d.* 2 *h.* 22 *m.*
		Scraped Bed. Alumina.			
July 27, "	4.00	2.30	2.15	30,000,000 (36,600,000)	0 *d.* 5 *h.* 28 *m.*
	The discharge-pipe during the remainder of the experiments was kept stationary until the water rose in the filter to a height of about 4.5 feet above the filter-bed, owing to the bed having become gradually clogged up. The discharge-pipe was then gradually lowered.				
		Scraped Bed. Alumina and "Free Flow."			
July 28, "	0.20	1.61	2.99	30,000,000 (22,200,000)	0 *d.* 1 *h.* 59 *m.*
		Scraped Bed. Alumina and "Free Flow."			
July 29, "	0.20	1.61	2.99	30,000,000 (29,700,000)	0 *d.* 2 *h.* 2 *m.*

* Small figures in parentheses show the average flow.

TABLE NO. 1.—CONTINUED.

FILTER No. 1.

Date of commencement of run.	Height of discharge-pipe in feet, above the filter-bed, at the commencement of the run.	Loss of Head in feet, due to the flow of the water through the filter-bed and discharge-pipe, at the commencement of each run.	The Equivalent Height in feet that the water rose in the filter during each run, after the filter commenced to discharge at its full capacity.	Gallons of Water Filtered per Acre, per 24 Hours, when the filter commenced to discharge at its full capacity, at the time that the loss of head was measured; also, the average flow of the filter during each run.*	Actual length of each run in days, hours, and minutes.
	Filter Bed Repacked with New Sand. Effective Size of the Sand Grains 0.35 mm. Uniformity coefficient 2.0. Natural Filtration.				
Aug. 1, 1893	0.20	0.25	4.20	30,000,000 (28,400,000)	2 *d.* 2 *h.* 12 *m.*
	Washed Bed with River Water. Alumina and "Free Flow."				
Aug. 3, "	0.20	0.70	3.77	30,000,000 (26,100,000)	0 *d.* 18 *h.* 27 *m.*
	Washed Bed with River Water. Natural Filtration.				
Aug. 5, "	0.20	0.36	5.92	30,000,000 (26,500,000)	4 *d.* 17 *h.* 29 *m.*
	Washed Bed with River Water. Natural Filtration.				
Aug. 10, "	0.20	0.37	5.93	30,000,000 (29,000,000)	4 *d.* 10 *h.* 47 *m.*
	Filter Bed Repacked with New Sand. Effective Size of the Sand Grains 0.81 mm. Uniformity coefficient 2.2. Alumina and "Free Flow."				
Aug. 16, "	0.20	0.39	5.91	30,000,000 (29,000,000)	5 *d.* 1 *h.* 8 *m.*

	Washed Bed with River Water. Alumina and "Free Flow."				
Aug. 21, 1893	0.20	0.37	5.93	30,000,000 (29,700,000)	3 *d.* 13 *h.* 35 *m.*
	Washed Bed with River Water. Alumina and "Free Flow."				
Aug. 25, "	0.20	0.48	4.78	30,000,000 (29,700,000)	1 *d.* 5 *h.* 41 *m.*
	Washed Bed with River Water. Alumina and "Free Flow."				
Aug. 28, "	0.50	0.45	3.65	30,000,000 (30,200,000)	2 *d.* 7 *h.* 31 *m.*
	Washed Bed with River Water. Alumina and "Free Flow."				
Aug. 31, "	0.50	0.65	3.45	30,000,000 (29,000,000)	1 *d.* 15 *h.* 29 *m.*
	Washed Bed with River Water. Alumina and "Free Flow."				
Sept. 2, "	0.50	0.56	5.54	30,000,000 (31,100,000)	1 *d.* 0 *h.* 12 *m.*
	Washed Bed with River Water. Alumina and "Free Flow."				
Sept. 4, "	0.50	0.49	3.59	30,000,000 (28,300,000)	1 *d.* 8 *h.* 32 *m.*
	Washed Bed with River Water. Alumina and "Free Flow."				
Sept. 6, "	0.50	0.53	3.54	30,000,000 (29,000,000)	2 *d.* 5 *h.* 42 *m.*

* Small figures in parentheses show the average flow.

TABLE NO. 1.—CONTINUED.

FILTER No. 1.

Date of commencement of run.	Height of discharge-pipe in feet, above the filter-bed, at the commencement of the run.	Loss of Head in feet, due to the flow of the water through the filter-bed and discharge-pipe, at the commencement of each run.	The Equivalent Height in feet that the water rose in the filter during each run, after the filter commenced to discharge at its full capacity.	Gallons of Water Filtered per Acre, per 24 Hours, when the filter commenced to discharge at its full capacity, at the time that the loss of head was measured; also, the average flow of the filter during each run.*	Actual length of each run in days, hours, and minutes.
	Washed Bed with River Water.	Alumina and "Free Flow."			
Sept. 8, 1893	0.50	0.50	3.59	30,000,000 (32,600,000)	2 *d.* 7 *h.* 47 *m.*
	Washed Bed with River Water.	Alumina and "Free Flow."			
Sept. 11, "	0.50	0.39	3.75	30,000,000 (30,300,000)	1 *d.* 3 *h.* 12 *m.*
	Washed Bed with River Water.	Alumina and "Free Flow."			
Sept. 12, "	0.50	0.33	3.76	30,000,000 (30,300,000)	1 *d.* 5 *h.* 22 *m.*
	Washed Bed with River Water.	Alumina and "Free Flow."			
Sept. 14, "	0.50	0.43	3.76	30,000,000 (31,100,000)	1 *d.* 6 *h.* 35 *m.*
	Washed Bed with River Water.	Alumina and "Free Flow."			
Sept. 27, "	2.00	0.38	0.48	30,000,000 (30,500,000)	1 *d.* 0 *h.* 0 *m.*

	Washed Bed with River Water.		Alumina and "Free Flow."		
Sept. 28, 1893..........	2.00	0.40	1.10	30,000,000 (28,900,000)	0 *d.* 21 *h.* 52 *m.*
	Washed Bed with River Water.		Alumina and "Free Flow."		
Sept. 29, "	2.00	0.56	0.96	30,000,000 (30,000,000)	0 *d.* 20 *h.* 41 *m.*
	Washed Bed with River Water.		Alumina and "Free Flow."		
Sept. 30, "	0.75	0.39	1.24	30,000,000 (27,600,000)	0 *d.* 20 *h.* 10 *m.*
	Washed Bed with River Water.		Alumina and "Free Flow."		
Oct. 2, "	0.50	0.55	2.31	30,000,000 (29,200,000)	1 *d.* 2 *h.* 50 *m.*
	Washed Bed with River Water.		Alumina and "Free Flow."		
Oct. 4, "	0.75	0.54	1.96	30,000,000 (30,500,000)	0 *d.* 23 *h.* 28 *m.*
	Washed Bed with River Water.		Alumina and "Free Flow."		
Oct. 5, "	0.75	0.45	2.21	30,000,000 (27,400,000)	1 *d.* 17 *h.* 20 *m.*

* Small figures in parentheses show the average flow.

TABLE No. 1.—CONTINUED.

Date of commencement of run.	FILTER No. 2. Height of discharge-pipe in feet, above the filter-bed, at the commencement of the run.	Loss of Head in feet, due to the flow of the water through the filter-bed and discharge-pipe, at the commencement of each run.	The Equivalent Height in feet that the water rose in the filter during each run, after the filter commenced to discharge at its full capacity.	Gallons of Water Filtered per Acre, per 24 Hours, when the filter commenced to discharge at its full capacity, at the time that the loss of head was measured; also, the average flow of the filter during each run.*	Actual length of each run in days, hours, and minutes.
	Discharge-pipe stationary. The water gradually rose in the filter as the bed clogged up. Effective Size of the Sand Grains 0.81 mm. Uniformity coefficient 2.2. Natural Filtration.				
Mar. 27, 1893	0.00	0.0104	0.0196	2,000,000 (1,709,000)	39 *d.* 17 *h.* 55 *m.*
	Water kept in filter at a constant height of about 4.0 feet above the bed. The discharge-pipe was gradually lowered as the bed clogged up. Filter Bed Repacked with Same Sand. Effective Size of the Sand Grains 0.81 mm. Uniformity coefficient 2.2. Natural Filtration.				
May 13, "	4.00	0.02	5.98	2,000,000 (1,830,000)	61 *d.* 4 *h.* 26 *m.*
	Washed Bed with Filtered Water. Natural Filtration.				
July 15, "	4.00	0.25	4.54	30,000,000 (27,800,000)	8 *d.* 15 *h.* 27 *m.*
	The discharge-pipe during the remainder of the experiments was kept stationary until the water gradually rose in the filter to a height of about 4.5 feet above the filter-bed, owing to the bed having become gradually clogged up. The discharge-pipe was then gradually lowered.				

	Washed Bed with Filtered Water. Alumina and "Free Flow."				
July 25, 1893..........	0.20	0.50	4.10	30,000,000 (17,300,000)	3 *d.* 0 *h.* 30 *m.*
	Washed Bed with Filtered Water. Alumina and "Free Flow."				
July 29, "	0.20	1.22	4.97	30,000,000 (28,500,000)	0 *d.* 7 *h.* 14 *m.*
	Alumina discontinued. Natural Filtration.				23 *d.* 21 *h.* 12 *m.*
	Scraped Bed. Natural Filtration.				
Aug. 23, "	0.50	0.39	3.96	30,000,000 (29,400,000)	5 *d.* 21 *h.* 4 *m.*
	Scraped Bed. Natural Filtration.				
Aug. 29, "	0.50	4.20	Did not reach its full capacity.	22,300,000	0 *d.* 2 *h.* 0 *m.*
	Scraped Bed. Natural Filtration.				
Aug. 30, "	0.50	1.02	3.07	30,000,000 (30,300,000)	4 *d.* 7 *h.* 33 *m.*
	Washed Bed with River Water. Natural Filtration.				
Sept. 4, "	0.50	0.27	3.86	30,000,000 (28,300,000)	9 *d.* 8 *h.* 39 *m.*
	Washed Bed with River Water. Natural Filtration.				
Sept. 27, "	2.00	0.32	1.94	30,000,000 (29,200,000)	7 *d.* 6 *h.* 6 *m.*
	Washed Bed with River Water. Natural Filtration.				
Oct. 5, "	0.75	0.33	3.34	30,000,000 (30,100,000)	10 *d.* 2 *h.* 57 *m.*

* Small figures in parentheses show the average flow.

TABLE NO. 1.—CONTINUED.

Date of commencement of run.	MORISON MECHANICAL FILTER.				
	Height of discharge-pipe in feet, above the filter-bed, at the commencement of the run.	Loss of Head in feet, due to the flow of the water through the filter-bed and discharge-pipe, at the commencement of each run.	The Equivalent Height in feet that the water rose in the filter during each run, after the filter commenced to discharge at its full capacity.	Gallons of Water Filtered per Acre, per 24 Hours, when the filter commenced to discharge at its full capacity, at the time that the loss of head was measured; also, the average flow of the filter during each run.*	Actual length of each run in days, hours, and minutes.
	Discharge-pipe stationary. The water gradually rose in the filter as the bed clogged up. Effective Size of the Quartz Grains 0.59 mm. Uniformity coefficient 1.5. Natural Filtration.				
April 5, 1893..........	3.75	2.00	3.42	128,000,000 (119,000,000)	4 *d.* 7 *h.* 20 *m.*
	Washed Bed with River Water. Natural Filtration after "Free Flow."				
May 11, "	3.75	2.44	3.56	128,000,000 (100,000,000)	3 *d.* 23 *h.* 35 *m.*
	Washed Bed with River Water. Natural Filtration.				
July 12, "	0.75	0.10	1.40	30,000,000 (18,500,000)	5 *d.* 20 *h.* 58 *m.*
	Washed Bed with River Water. Alumina and "Free Flow."				
Aug. 21, "	3.75	0.45	5.60	30,000,000 (29,000,000)	2 *d.* 7 *h.* 21 *m.*
	Washed Bed with River Water. Alumina and "Free Flow."				
From April 17 to July 8, and from Aug. 8 to Sept. 11, 1893.	† 3.75	2.44	3.56	128,000,000 (115,000,000)	Average. 0 *d.* 16 *h.* 38 *m.*
	Washed Bed with River Water. Alumina and "Free Flow."				
From July 10 to Aug. 7, 1893.	0.75	2.44	3.56	128,000,000 (121,000,000)	Average. 0 *d.* 17 *h.* 38 *m.*

† Includes runs when "Free Flow" was not used.

* Small figures in parentheses show the average flow.

TABLE NO. 1.—CONTINUED.

Average Color of the Water during each run, of Filters Nos. 1 and 2.

(Did not begin to observe color until May 20, 1893.)

FILTER No. 1.				FILTER No. 2.			
Date run commenced.	Average color of Applied Water.	Average color of Filtered Water.	Per cent. of color removed.	Date run commenced.	Average color of Applied Water.	Average color of Filtered Water.	Per cent. of color removed.
May 13, 1893....	....	6.0	....	May 13, 1893.....	....	6.6	....
June 15, "	.. .	7.1	....	July 15, "	....	8.8	....
July 17, "	....	9.0	....	" 25, "	+10.0	6.1	39.
" 26, "	+10.0	1.0	90.	" 29, "	+10.0	9.8	2.
" 27, "	+10.0	7.7	23.	Aug. 23, "	4.5	3.0	33.
" 28, "	+10.0	1.5	85.	" 29, "	4.5	3.0	33.
" 29, "	+10.0	1.3	87.	" 30, "	4.5	3.0	33.
Aug. 1, "	+10.0	10.0	00.	Sept. 4, "	4.6	2.8	39.
" 3, "	+10.0	2.5	75.	" 27, "	4.3	3.3	20.
" 4, "	+10.0	8.9	11.	Oct. 4, "	4.6	2.9	37.
" 5, "	+10.0	10.0	00.				
" 10, "	+10.0	10.0	00.				
" 16, "	+10.0	8.9	11.				
" 21, "	5.0	2.8	44.				
" 25, "	4.0	1.1	73.				
" 28, "	4.7	1.6	66.				
" 31, "	5.0	1.0	80.				
Sept. 2, "	5.0	0.7	86.				
" 4, "	5.0	0.7	86.				

TABLE NO. 1.—CONTINUED.

FILTER No. 1.

Date run commenced.	Average color of Applied Water.	Average color of Filtered Water.	Per cent. of color removed.
Sept. 6, 1893....	4.7	2.4	49.
" 8, "	5.0	1.2	76.
" 11, "	4.0	0.9	78.
" 12, "	4.0	0.8	80.
" 14, "	4.0	0.7	83.
" 27, "	4.0	1.7	58.
" 28, "	4.0	1.9	52.
" 29, "	5.0	2.2	56.
" 30, "	4.0	1.5	63.
Oct. 2, "	4.3	1.7	60.
" 4, "	5.0	1.6	68.
" 5, "	4.0	1.8	55.

The average percentage of color removed from the water by filtration through the Morison Mechanical Filter, when Basic Sulphate of Alumina was used, is: From June 28, to Oct. 26, 1893, 81.0, and from Oct. 28, 1893, to Jan. 30, 1894, 78.0.

TABLE NO. 1.—CONCLUDED.

Quantity of Basic Sulphate of Alumina Used.

During the periods covered by the table when Basic Sulphate of Alumina was used, the average rate added to the Applied Water per gallon was :—

IN THE MORISON MECHANICAL FILTER.

On April 29, $\frac{31}{100}$ of a grain, including "Free Flow"; on June 16, $\frac{26}{100}$ of a grain, not including "Free Flow"; and on June 27, $\frac{35}{100}$ of a grain, including "Free Flow." From April 14 to April 25 inclusive, from June 1 to June 3 inclusive, and on June 6 and July 6, $\frac{50}{100}$ of a grain, not including "Free Flow," and for all other periods during the experiments with this filter $\frac{60}{100}$ of a grain, including "Free Flow."

IN FILTER NO. 1.

From July 26 to July 27 inclusive, $\frac{50}{100}$ of a grain, not including "Free Flow." On October 4, $1\frac{11}{100}$ grains, including "Free Flow," and for all other periods during the experiments with this filter $\frac{60}{100}$ of a grain, including "Free Flow."

IN FILTER NO. 2.

From July 25 to July 29 inclusive, $\frac{60}{100}$ of a grain, including "Free Flow."

Description of the Experiments that were made with the Experimental Morison Mechanical Filter.

There are a number of cases throughout this report, in the text and tables, where the Experimental Morison Mechanical Filter is called the Morison Mechanical Filter, the latter, however, is simply an abbreviation of the former.

A sketch representing the Experimental Morison Mechanical Filter that was used during the filtration experiments is shown in Cut No. 2.

Upon the screens shown at the bottom of the filter, the filtering medium, or filter-bed, of crushed quartz is located, the total depth being two (2) feet and ten (10) inches. The Effective Size of the grains of quartz which compose the upper two (2) feet is 0.59 millimeters, and the Uniformity Coefficient 1.5. The lower ten (10) inches of quartz is of a much coarser quality. The screens allow the water to pass through them during the different operations of working the filter, downward while filtering, and upward during the process of washing the filter-bed. They prevent the quartz or any foreign substances from entering the collecting pipes, or passing off with the filtered water.

The manner in which the filter was operated during the experiments is as follows: At the end of a run, or immediately before starting the filter, the filter-bed was thoroughly washed by forcing up through the screens and filter-bed a reverse flow of water under pressure, the mechanical rake or agitator shown in the cut being operated at the same time, which added materially to the efficient cleansing of the filter-bed. The water was forced up through the bed and the agitator kept in motion until the water flowing from the overflow drain-pipe was as clear as it was before it was used for washing the filter-bed. The necessary valves were then operated and the water and Sulphate of Alumina turned on to the filter.

The rates of the filtration of water, mentioned in this report, all represent an average rate per Acre per 24 hours, unless otherwise specified. The standard rate of filtration decided upon at the commencement of the experiments was 128,000,000 gallons per Acre per 24 hours. When the term Sulphate of Alumina is used it is intended as an abbreviation of Basic Sulphate of Alumina.

Cut No. 2.

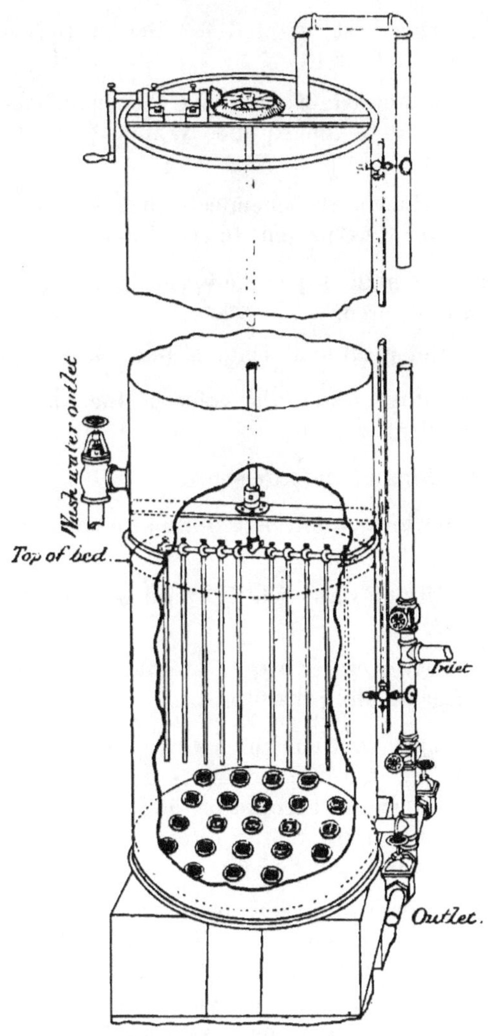

Morison Experimental Filter.

Height 14 feet, Diameter 30 inches

In making the experiments with this filter, the following details were carefully investigated, as well as many other points relative to the efficient working of the filter, viz :—

First.—The chemicals best adapted for the purification of the Pawtuxet River Water.

Second.—The best method of applying the chemicals, and the quantity to add to the Applied Water for each gallon of water filtered.

Third.—If any portion of the chemicals that were added to the Applied Water were present in the Filtered Water.

Fourth.—The rate in gallons per Acre per 24 hours, which could be efficiently filtered.

Fifth.—The bacteriological and chemical purification of the water.

Sixth.—The percentage which the color of the water would be reduced by filtration.

Seventh.—The washing of the filter-bed.

Eighth.—The time which would be required for washing the filter-bed.

Ninth.—The quantity of water which would be required to wash the filter-bed.

Tenth.—The quantity of water which it would be necessary to run to waste after washing the filter-bed.

Eleventh.—The length of time which the filter would run after starting, before it would be necessary to shut down and wash the filter-bed on account of the water gradually rising to its prescribed limit in the filter, owing to the filter-bed becoming gradually clogged up.

Twelfth.—The effective stability of the quartz and supplementary precipitate bed, viz :—whether it could be depended upon to do its work thoroughly during the whole of the time that the filter was in operation or whether at times it would be liable to crack or break, or have its efficiency reduced in any manner.

Thirteenth.—The loss of head due to the water flowing through the filter-bed and screens.

During the preliminary experiments, the chemicals used were Basic Sulphate of Alumina, Chloride of Alumina, Carbonate of Soda, Bicarbonate of Soda, Caustic Soda, and Chloride of Iron. The soda salts were used in connection with Sulphate of Alumina. It was found, however, that Basic Sulphate of Alumina added to the Applied Water produced the best results. Basic Sulphate of Alumina, therefore, is the only chemical that has been used since the preliminary experiments.

The theory of Mechanical Filtration, when Basic Sulphate of Alumina has been added to the Applied Water, may be described as follows: The Alumina causes an artificial precipitation. A portion of the Alumina is decomposed, forming sulphates of other bases and a flocculent precipitate of Aluminic Hydrate. A portion of it also combines directly with the organic matter present in the water, coagulating the same and thus helping to increase the precipitation. A sufficient quantity of the precipitate having been deposited upon the top of the sand or quartz-bed of the filter and plugged into the interstices of the upper layer of sand or quartz grains, the filter is ready for service.

At the commencement of the experiments with the Morison Mechanical Filter, it was discovered, that satisfactory results could not be obtained by simply dropping the Sulphate of Alumina into the Applied Water at the rate of one-half (½) grain per gallon, as it would take from 1 to 3 hours after the filter was started for a sufficient quantity of the precipitate to form in order to do good work. After experimenting in different ways, it was found, that if a "Free Flow" of about a pint of coagulant containing about nine hundred and eleven (911) grains of Sulphate of Alumina, for an average rate of filtration of about 128,000,000 gallons per Acre per 24 hours, was allowed to run into the filter, immediately after the water was let on, in a space of time of not more than six (6) minutes, a quantity of coagulant corresponding to one-half (½) grain of Sulphate of Alumina per gallon of Filtered Water being dropped in at the same time from a different receptacle than that containing the "Free Flow," a sufficient amount of precipitate would be formed to do good work in one-half hour or less after the water commenced to flow from the filter.

At the commencement of a run of the filter, the Applied Water, was, at first, gradually let into the filter, it being regulated at the same time. After the normal quantity commenced to flow into the filter a constant flow was maintained, and the depth of water in the filter gradually increased proportionately during the run as the supplementary precipitate bed was formed and the filter-bed became plugged with precipitate. The rise of water practically accommodated itself to the circumstances, and caused a constant flow of water through the filter, which I considered extremely essential in order to obtain good results.

Two experiments, that were made about five months after the filter was first started for the purpose of ascertaining the loss of head due to the addition of the "Free Flow" and Sulphate of Alumina as above described, demonstrated, that the addition of the "Free Flow" and Sulphate of Alumina caused a loss of head of about twenty-eight hundredths (0.28) of a foot (for a rate of 128,000,000 gallons), in addition to the loss of head which was caused by the passage of the water through the quartz-bed. These two experiments each covered a space of time of about thirty-five (35) minutes, from the time that the water was first let on to the filter until a rate of 128,000,000 gallons per Acre per 24 hours was reached.

A number of experiments that have been made during the progress of the work, by running the filter as a "natural filter," have shown, approximately, that the average efficiency of the filter, for removing water bacteria, was increased about thirty-eight (38) per cent., while filtering at an average rate of about 30,000,000 and 127,000,000 gallons, by adding the "Free Flow" and Sulphate of Alumina at the rate of one-half (½) grain per gallon.

A description of the method that was followed in adding the Basic Sulphate of Alumina to the Applied Water is as follows: The coagulant used (including the "Free Flow"), was made by dissolving one (1) part of the Sulphate of Alumina (by weight), in seven (7) parts of water (by weight). The cisterns that were used in adding the coagulant to the Applied Water were located upon the top of the filter. The cistern that supplied the "Free Flow," was simply a small tin can with a faucet, which could be regulated as desired. The cistern that fed the Sulphate of Alumina, at the rate of one-half (½) grain or more per gallon, was an earthern jar which had a small faucet connected near its bottom. To the outlet of this faucet was connected a small glass dropper. The faucet was so regulated, by the aid of a carefully graduated measuring glass, that the coagulant would drop from the "dropper" into the filter at an average rate of about sixty (60) drops per minute (for a rate of one-half (½) grain of Sulphate of Alumina per gallon), for a rate of filtration of 128,000,000 gallons per Acre per 24 hours. These drops were counted every half hour, day and night, while the filter was running, and the faucet regulated when necessary. The Sulphate of Alumina coagulant and "Free Flow" were turned on immediately after the water commenced to flow into the filter, at which time there was always a depth of about nine (9) inches or more of water upon the quartz-bed. The drops of coagulant fell into the interior of the filter upon the surface of the water and were thoroughly mixed with the water by the agitation produced by the water falling from the outlet of the supply-pipe.

One of the most serious problems that it was necessary to solve when the experiments were commenced, was to ascertain if the Basic Sulphate of Alumina, that was added to the Applied Water, was entirely decomposed before the water was discharged from the filter.

I was informed by two eminent chemists, that it would be a very difficult matter, to positively ascertain the quantity of Alumina, if there was any, in the Applied or the Filtered Water, although the Aluminum Compounds could be determined without much difficulty. It was decided, therefore, to make use, principally, of the "Logwood and Acetic Acid test," for alum, of Mrs. E. H. Richards of the Massachusetts Institute of Technology. "Chemical Analyses" were also made, by Professors Appleton and Drown, as will be seen by the tables and appendix.

It was found, during preliminary experiments with the Logwood and Acetic Acid test, by adding different quantities of Sulphate of Alumina to Distilled Water, and by exercising great care and using fresh Logwood decoction carefully prepared, that one (1) part of Sulphate of Alumina in 1,000,000 parts of water could be detected, which is equivalent to about one (1) part of Alumina (Al_2O_3) in 6,000,000 parts of water or $\frac{1}{100}$ of a grain to a gallon of water.

A sample of the Sulphate of Alumina used had the following composition :—

	Per cent.	One-half (½) grain contains in grains.
Insoluble residue........................	0.52	0.0026
Alumina (Al_2O_3).	15.78	0.0789
Sulphur trioxide (SO_3)...	36.79	0.1840
Water (by difference)....................	46.91	0.2345
	100.00	0.5000

The Logwood and Acetic Acid test was applied to the Filtered Water for a number of days, and in nearly every instance the test indicated the presence of Alumina in solution. This of course was a very serious matter, as the presence in the Filtered Water of Alumina in solution as the test seemed to indicate, was a strong argument against mechanical filtration. I consulted with Mrs. Richards in regard to the application of the Logwood and Acetic Acid test, as well as with several other chemists in regard to the matter, but they were not able to throw any new light upon the subject. Finally Dr. S. C. Hooker, a chemist of Philadelphia, suggested to Dr. Chapin, Superintendent of Health, who was also investigating the subject, that the alum tint, produced in the Filtered Water by Mrs. Richards' test, might be due to a small quantity of finely suspended hydrate, which could be proved by a careful filtration of the Filtered Water through filter-paper and then applying the Logwood and Acetic Acid test.

A short time after Dr. Hooker suggested this method of treating the Filtered Water, it was applied in my presence by Professor C. A. Doremus in his New York laboratory.

Two samples of equal quantity, of Filtered Water from the Morison Mechanical Filter, were taken from the flask containing the same ("Free Flow" and one-half (½) grain of Sulphate of Alumina per gallon having been added to the water before filtration). One sample was filtered through two thicknesses of fine German filter-paper, the Logwood and Acetic Acid test was then applied to each sample, and the alum tint was produced in the sample that had not been filtered through the paper, while the sample that had been filtered through the paper was entirely free from it. After my return to Providence, I made quite a number of tests in the manner above described with Filtered Water and with River Water to which Basic Sulphate of Alumina had been added at the rate of one-half (½), three-fourths (¾), and one (1) grain per gallon, and obtained the same results, namely:—no traces of the alum tint were detected after the application of the Logwood and Acetic Acid test in any of the samples of water that had been filtered through paper. Several tests were also made with Distilled Water freshly distilled from River Water, to which Sulphate of Alumina had been added at the rate of one-half (½) grain per gallon, and the alum tint was visible both before and after filtration, though of a slightly darker shade in the former. The alum tint was not produced in the Distilled Water owing to the absence of the constituents from the Distilled Water necessary to decompose the Alumina and form a hydrate, consequently it passed through the filter-paper, in the Distilled Water, in a soluble form.

While we were in doubt as to the complete decomposition of the applied Alumina (before Dr. Hooker's suggestion, that the alum tint produced by the Logwood and Acetic Acid test in the Filtered Water was due to a finely suspended hydrate instead of Alumina in solution, had been found to be correct), several experiments were made with two settling tanks (having a combined area about ten times the area of the filter), in order to ascertain, if possible, if a more complete chemical action would take place, if a longer length of time was allowed to elapse before the Applied Water (after the Sulphate of Alumina had been added), reached the filter-bed. During these experiments the Applied Water was first run into a tank (having an area of about sixteen (16) square feet, and a depth of two and twenty-five hundreths (2.25) feet). The Sulphate of Alumina coagulant was dropped into this tank instead of being dropped directly into the filter, and the "Free Flow" put into the filter in the usual way. The water flowed from this tank, through an orifice, located about four (4) inches above its bottom, into a larger tank, situated directly under it (having an area of about thirty-three (33) square feet, and a depth of two and twenty-five hundreths (2.25) feet), and from this latter tank, through a pipe, one and one-quarter (1¼) inches in diameter, connected about three (3) inches above its bottom, into the filter. It took the Applied Water, which flowed continually through both tanks, about twenty-two (22) minutes to pass through the first tank and about fifty-three (53) minutes to pass through the second tank. The results obtained, from the experiments that were made with the settling tanks, were not quite as satisfactory, from a bacteriological standpoint, as the results obtained by dropping both the Sulphate of Alumina coagulant and "Free Flow" into the filter, in the usual way, as has already been described, and there was not any diminution in the indications of Alumina in solution in the Filtered Water, so far as could be discovered by applying the Logwood and Acetic Acid test.

When the filter was started the water commenced to flow from the outlet-pipe, generally, about five (5) minutes after it was turned on to the filter. A sample of the Filtered Water was always collected one (1) minute after it commenced to flow, and five samples, one every five (5) minutes for one-half hour, and then hourly during the day and several times during the night. No alum tint was ever visible in the one (1) minute and six (6) minute samples, when the Logwood and Acetic Acid test was applied, either before or after filtration through paper. In all of the other samples, the alum tint was visible before filtration through paper, and in the eleven (11) and sixteen (16) minute samples it was visible after filtration through paper and occasionally in the twenty-one (21) minute sample, but it never was detected in any of the samples taken later than twenty-one (21) minutes after the water commenced to flow from the filter. The eleven (11), sixteen (16), twenty-one (21) and twenty-six (26) minute samples had generally, before filtration through paper, a darker tint, which grew less as the time increased, than the hourly and night samples. The same may be said of the eleven (11) and sixteen (16) minute samples after filtration through paper, and of the twenty-one (21) minute sample when the alum tint was visible. Great care was taken to have the Logwood decoction prepared properly. It was also necessary to guard against filter-paper that contained traces of aluminum salts. The paper was always tested by applying the Logwood and Acetic Acid test to two samples of freshly Distilled Water, one of which had been filtered through the paper, and one of which had not. The best results are obtained with the Logwood and Acetic Acid test, when it is applied expeditiously, and in making the test, in order to aid in detecting the alum tint, it should be applied to a sample of Distilled Water at the same time that it is applied to the samples of Filtered Water.

Bacteriological Work.

As I was informed when the filtration work was first commenced, that the bacteriological cultivations and counts of the samples of water would be made under the direction of the Superintendent of Health, I did not assume any direct responsibility in regard to the bacteriological work until early in September. At this time I commenced to personally investigate the subject, and was asked by the Superintendent of Health to make any suggestions that I deemed advisable and to give such directions as I thought proper in regard to this work. After making myself familiar with the methods that were generally followed in this country and abroad, and consulting with Professor H. C. Ernst of the Harvard University Medical School, who had made some test counts for us upon two occasions, I came to the conclusion that the bacteria had not been cultivated long enough to reach their highest growths, and that the "Fifteen-per-cent Gelatin" that had been used, nearly the whole of the time, should be discontinued and "Ten-per-cent Gelatin" used, which is the nutrient media generally made use of in the cultivation of water bacteria.

The majority of the counts from March 27 to October 6, 1893, had been made after a cultivation of from 42 to 62 hours. Subsequent investigations have proved that these counts were made too soon, and that more bacterial colonies would have been visible when the counts were made, if a longer period had been allowed for cultivation and "Ten-per-cent Gelatin" used. I am, therefore, of the opinion that the bacteriological counts that were made from March 27 to October 6, 1893, are not strictly reliable, and should only be used for comparing the efficiency of the different filters. The application, however, of a slight correction, derived by comparing the counts of from 42 to 62 hours with results that have been obtained since October 6, 1893, tends to show, when Basic Sulphate of Alumina and "Free Flow" were used, that an average of about ninety-nine (99) per cent. of the Applied Water bacteria was removed by the Morison Mechanical Filter, from March 27 to October 6, 1893. This average percentage, if it had not been corrected, would have been slightly more than ninety-nine per cent.

Owing to the reasons given above, the only bacteriological work which I shall mention, and describe hereafter, unless otherwise specified, will be the work that was done later than October 6, 1893, or that which was done previous to that time by Professor H. C. Ernst.

The method used in cultivating the bacterial colonies was the familiar method of gelatin-plate culture devised by Koch.

The bacterial colonies were grown at the laboratory temperature.

The bacterial colonies visible in each dish were first counted after a cultivation of about 42 hours, and subsequently about every 24 hours until an increase in their number could no longer be detected. It was then assumed that their end growths had been reached. The entire length of time necessary for cultivation ranged from 67 to 236 hours.

A great deal of trouble was caused by the bacterial colonies liquefying before end growths were reached (from October 17,

1893, to January 30, 1894, all of 51 days' samples were lost on this account), and it was deemed advisable on December 12, 1893, to discontinue using one (1) cubic centimeter from each sample of Filtered Water which had previously been used in each dish and to use four dishes for each sample with one-fourth ($\frac{1}{4}$) of one (1) c.c. in each dish.

This method, though being an improvement upon the use of one (1) c.c., was not entirely satisfactory, and on December 26, 1893, another change was made, namely:—the equal division, from each sample, of one-half ($\frac{1}{2}$) of one (1) c.c. among five dishes. This latter method was followed until the completion of the work, and very little inconvenience was experienced from liquefying colonies during this time. A check was kept upon it by frequently cultivating one (1) c.c. in one dish, from the same sample that one-half ($\frac{1}{2}$) of one (1) c.c. divided by five (5) was taken from, and the average result obtained from all the one (1) c.c. cultures, which could be kept 137 hours or more without liquefying, was almost exactly the same as the average of the one-half ($\frac{1}{2}$) of one (1) c.c. divided by five (5) cultures that were taken from the same samples.

Four dishes, each containing $\frac{1}{100}$ of one (1) c.c. of Applied Water, were used in making the cultures of the Applied Water, with the exceptions which will be mentioned hereafter.

The methods of cultivating the samples of Filtered Water, when Bacillus Prodigiosus was being added to the Applied Water, were the same as those that have previously been described; but only in a few instances was it possible to ascertain, on account of liquefaction, if end growths had been reached. But as it is generally customary to count cultures of these bacilli after from 48 to 96 hours growth, and as table No. 19, giving the results that were obtained when they were being applied, shows that the samples of Filtered Water were cultivated from 41 to 208 hours before they liquefied, and the samples of Applied Water from 40 to 190 hours, there is not any doubt, I think, but what the percentages given in this table are sufficiently reliable.

A pure culture of Bacillus Prodigiosus was obtained, by inoculation and growth for about four days, in the following solutions, namely: On November 22, 23 and 25, 1893, one (1) liter of Bouillon; on the 28 and 29, four (4) cubic centimeters of the above Bouillon in one (1) liter of tap water; on December 2, 5 and 6, a solution of one-tenth ($\frac{1}{10}$) per cent. Peptone, and two-tenths ($\frac{2}{10}$)

per cent. Glucose in tap water; on the 12, 13 and 14, four (4) cubic centimeters of the above Peptone and Glucose solution in one (1) liter of Sterile Water; on the 15 and 16, five (5) cubic centimeters of the Peptone and Glucose solution in one (1) liter of Sterile Water; on the 18 and 19, four (4) cubic centimeters from one (1) liter of Bouillon and "Ten-per-cent Gelatin" in which the culture was made, in one (1) liter of Sterile Water; on the 20, 21 and 22, the full Peptone and Glucose solution; from December 3, 1893, to January 3, 1894, inclusive, one-tenth ($\frac{1}{10}$) of the Peptone and Glucose solution in one (1) liter of tap water; and from January 4 to 8, one-twentieth ($\frac{1}{20}$) of the Peptone and Glucose solution in one (1) liter of tap water.

The Bacillus Prodigiosus Solution was uniformly applied by being dropped into the filter from an earthen jar located upon the top of the filter.

Cruikshank's Bacillus, that was prepared and used in a manner similar to the Bacillus Prodigiosus, was added to the Applied Water on July 27 and August 17, and on October 11 and 12, 1893, at the rate of more than one million (1,000,000) per cubic centimeter. Three (3) colonies of this bacillus were found, after a cultivation of five days, in the sample of Filtered Water that was collected on July 27. None were discovered in the samples of Filtered Water that were collected on the other days mentioned.

The samples of Applied Water containing the Bacillus Prodigiosus were drawn from a stop-cock connected to the filter about seven (7) inches above the filter-bed. Samples of the Applied Solution were also taken during each run, and if the samples from the top of the filter liquefied in cultivation, the number of Bacillus Prodigiosus in the Applied Water were estimated from the number in the Applied Solution.

The Bacillus Prodigiosus Applied Water was generally cultivated in four dishes, two, each containing $\frac{1}{1000}$ c.c. and two, each containing $\frac{1}{10000}$ c.c. The Applied Solution was cultivated in two dishes, each generally containing $\frac{1}{1000000}$ c.c.

The proportions of $\frac{1}{100}$ c.c. of the Applied Water and the $\frac{1}{1000}$ c.c. and $\frac{1}{10000}$ c.c. of the Bacillus Prodigiosus Applied Water and the $\frac{1}{1000000}$ c.c. of the Applied Solution, were obtained, for cultivation, by diluting with Sterile Water of known volume. The proportions of one-fourth ($\frac{1}{4}$) of one (1) c.c. and one-tenth ($\frac{1}{10}$) of one (1) c.c. of the Filtered Water were obtained by direct measurement without dilution.

The samples of Applied Water (not containing Bacillus Prodigiosus), were collected from a tap in the pipe which supplied the filter with water. All samples of Filtered Water were collected at the outlet of the discharge-pipe of the filter.

The gelatin used in the bacteriological work was tested for alkalinity up to about the first of November with litmus paper and after that time by the "phènol-phthalèin test."

The following table, computed from results obtained in our laboratory, shows the Ratios of the growths of bacterial colonies in "Ten-per-cent Gelatin," over what they were in "Fifteen-per-cent Gelatin," etc.

NUMBER OF HOURS OF CULTIVATION, ETC.

Number of Samples.	41 ± Hours.	Number of Samples.	67 ± Hours.	Number of Samples.	89 ± Hours.	Number of Samples.	113 ± Hours.	Number of Samples.	137 ± Hours.	Date, December, 1893.
3	1.08	3	1.36	3	1.58	2	1.09	1	0.94	14
10	1.06	10	1.14	10	1.01	9	0.85	7	0.79	15
4	3.00	4	1.16	4	1.17	3	1.28	..	....	16
1	1.53	1	*1.31	....	....	....	....			18
16	1.23	8	1.18	...	....	...	....	...		19
16	1.33	8	1.41	....	....	....		...		20
18	1.89	18	1.69	12	1.62	....	....	...		21
Average ratios.	1.60	...	1.32	....	1.35	....	1.07	...	0.87	...

* Not included in averages.

The above table indicates that nearly the same results can be obtained with "Fifteen-per-cent Gelatin" as can be obtained with

"Ten-per-cent Gelatin," if the bacteria are grown more than 113 hours.

The Bacteriological Work, in the laboratory was done from March 27, to December 10, 1893, by Dr. G. T. Swarts, Medical Inspector, with the exception of several short periods of time, when it was done by Dr. C. V. Chapin, Superintendent of Health. From December 11, 1893, to January 31, 1894, the work was done by Dr. M. T. Richardson, a graduate of the Harvard University Medical School, who was recommended by Professor H. C. Ernst. The remainder of the time, from February 1, to February 12, the counts were made by myself.

Since October 1, I have had consultations in regard to the Bacteriological Work, with Professor H. C. Ernst of the Harvard University Medical School, Professor E. K. Dunham of the Carnegie Laboratory of New York, and Professor T. M. Prudden of Columbia College. These gentlemen, all of whom are expert bacteriologists, have signified their approval of the methods that have been followed in regard to the Bacteriological Work, since October 6, 1893.

Description of Tables.

All, of the tables from No. 2 to No. 21 inclusive refer to the Experimental Morison Mechanical Filter.

The columns in the Bacteriological tables from No. 2 to No. 8 inclusive headed "Grains of Sulphate of Alumina used per Gallon," include "Free Flow." The rate of Alumina (not including "Free Flow"), that was added constantly to the Applied Water, was gauged very carefully and it was intended to apply it at a specific rate per gallon, as the case might require, of one-half ($\frac{1}{2}$) grain or three-fourths ($\frac{3}{4}$) of a grain (or more or less). This could not always be done in actual practice, however, as the outlet of the "dropper" was so small that it would sometimes clog up in spite of constant watchfulness and testing. The computed average rate, including "Free Flow" and a constant addition of Alumina at the rate of one-half ($\frac{1}{2}$) grain per gallon, would be fifty-nine one-hundreths ($\frac{59}{100}$) of a grain for an average length of run of 16 hours and 43 minutes and an average rate of filtration of 128,000,000 gallons per Acre per 24 hours, and under the same conditions the computed average rate, including "Free Flow" and a constant addition of three-fourths ($\frac{3}{4}$) of a grain of Alumina per gallon, would be eighty-four one-hundredths ($\frac{84}{100}$) of a grain. The quantity of "Free Flow" used was always the same, therefore an increase or decrease in the length of the run would of course change the computed average rate of Alumina used during the entire run, and a slight deviation one way or the other in the quantity of water filtered would also change the average rate of Alumina used. More than the usual care was exercised in all gaugings of both Alumina and Water for at least one-half hour before the samples of Filtered Water were collected.

Tables from No. 2 to No. 19 inclusive give the Bacteriological Results that have been obtained and the percentages of the Bacteria in the Applied Water that were removed by filtration, computed from the same. Only one sample of Applied Water was generally collected each day, the hour of collection in the great majority of cases being from 12 M. to 1 P. M. The samples of Filtered Water were collected as will be seen from the tables, from One (1) to Ten (10) times daily.

Table No. 2 gives the End Growths or positive results that were obtained from Samples of Filtered Water that were collected at the Same Hour as the Applied Water, once during each run, generally from 12 M. to 1 P. M., One Hour or More after water commenced to flow from the filter.

Table No. 3 covers the same ground as table No. 2, with the exception that it is computed from counts that were made after a cultivation of about Ninety Hours. This table was made for comparison with table No. 2, as it is generally customary to make the counts of water bacteria after they have been cultivated about four days.

Table No. 4 gives the End Growths or positive results that were obtained from Samples of Filtered Water that were collected from One (1) to Nine (9) times each day, One Hour or More after water commenced to flow from the filter. The percentages were all based upon results derived from the single sample of Applied Water that was generally collected from 12 M. to 1 P. M. daily.

Table No. 5 gives the End Growths and all of the Growths of Eighty Five Hours or More, that did not reach their End Growths, that were obtained from Samples of Filtered Water that were collected from One (1) to Nine (9) times each day, One Hour or More after water commenced to flow from the filter. The percentages were all based upon results derived from the single sample of Applied Water that was generally collected from 12 M. to 1 P. M. daily.

Table No. 6 gives the End Growths or positive results that were obtained from Samples of Filtered Water that were collected Thirty Minutes or Less after water commenced to flow from the filter. The percentages were all based upon results derived from the single sample of Applied Water that was generally collected from 12 M. to 1 P. M. daily. River Water was used in washing the filter, with the exception of on November 15, 17, 18, 20, 23 and 24, when Filtered Water was used.

Table No. 7 covers the same ground as table No. 6, with the exception that it was computed from counts that were made after a cultivation of about Ninety Hours. This table was made for comparison with table No. 6, for the same purpose as is mentioned in the description of table No. 3.

Table No. 8 gives the End Growths and all of the Growths of Eighty Five Hours or More, that did not reach their End Growths, that were obtained from Samples of Filtered Water that were collected Thirty Minutes or Less after water commenced to flow from the filter. The percentages were all based upon results derived from the single sample of Applied Water that was generally collected from 12 M. to 1 P. M. daily.

In tables Nos. 2, 3, 4, 5, 6, 7 and 8 during the time that Bacillus Prodigiosus was used, the number of bacteria in the columns headed "In Filtered Water," include the Bacillus Prodigiosus, when there were any, found in the Filtered Water. (See table No. 19). If they had not been included, the average per cents, of the "Applied Bacteria Removed," would be slightly larger and a few of the individual per cents considerably larger. This does not affect, however, any of the final results or conclusions which will be mentioned hereafter in this report, for reasons which will be subsequently explained.

The average percentages given in the tables from No. 2 to No.

8 inclusive, which are not inclosed in parentheses, and which were considered as a basis for all comparisons and summaries, were obtained by averaging the individual per cents given in the tables. The average percentages obtained by using the total number of bacteria found in the Applied and Filtered Water are also given in the tables, inclosed in parentheses, in order to show the difference between this method of computation, which is sometimes followed for obtaining average bacterial percentages, and the method just previously mentioned. The average percentages obtained by using total numbers, as can be seen by inspecting the tables range from 0.0 to 3.1 more than the averages obtained by using the individual per cents of each sample.

Tables Nos. 9 and 10 were computed from tables Nos. 2, 3, 4, 5, 6, 7 and 8, and give Summaries of the Average Percentages of Applied Water Bacteria that were Removed by the filter. The averages obtained by "totals" are also given in parentheses in these two summaries. They were obtained in the same manner as described above for tables from No. 2 to No. 8 inclusive.

Tables from No. 11 to No. 18 inclusive, were computed from tables Nos. 2, 3, 4, 5, 6, 7 and 8.

Tables Nos. 11 and 15 give the number of times that Percentages of More than Two Per cent., of the Applied Water Bacteria, Appeared in the Filtered Water. Also the Percentages that the number of times are of the Total Number of Results obtained.

Tables Nos. 12, 14, 16 and 18 give the number of times that Percentages of the Applied Water Bacteria Removed, which were used in working up the Average Percentages given in tables Nos. 9 and 10, were One Per cent. and More Less than the average Per cent. Removed. Also the Percentages that the number of times are of the Total Number of Results obtained.

Tables Nos. 13 and 17 give the number of times that Percentages of the Applied Water Bacteria Removed, which were used in working up the Average Percentages given in tables Nos. 9 and 10, were More than Two Per cent. Less than the Average Per cent. Removed. Also the Percentages that the number of times are of the Total Number of Results obtained.

Table No. 19 gives the Percentage of Applied Bacillus Prodigiosus that was Removed from the water by filtration. Also the number of these bacilli that were found in the Applied and Filtered Water and the length of time that they were Grown. The "Last Growth obtained," mentioned in the table, was the last growth that could be obtained before the bacterial colonies liquefied. The Average Percentages given in the table were obtained in the same manner as the averages given in tables from No. 2 to No. 8 inclusive. The quantity of Alumina used per gallon of Applied Water is given in the latter part of the table.

Table No. 20 gives the Chemical Analyses of Applied and Filtered Water that were made during the experiments with the Experimental Morison Mechanical Filter, by Professor J. H. Appleton.

Table No. 21 gives the Color of samples of Applied and Filtered Water that were collected during the experiments with the Experimental Morison Mechanical Filter, and the Percentage of Color that was Removed from the Applied Water by filtration.

Table No. 22. As the elaborate and very valuable experiments relative to the Natural Filtration of water, that have been made at Lawrence, Massachusetts, under the direction of the State Board of Health of Massachusetts, during the past few years, are recognized, I think, by the engineering profession the world over, as being the most complete exposition of the subject that has ever been made, table No. 22 has been compiled from the Report of the State Board of Health of Massachusetts for the year 1892, in order to make, in a few instances, a general comparison of some of the results that have been obtained at Providence with the Morison Mechanical Filter with some of the results that have been obtained by Natural Filtration with Experimental Filters at Lawrence.

The Massachusetts Report states, relative to some of the data that has been used in computing the table, namely: "It has" "been found, however, that the true degree of bacterial purifica-" "tion is somewhat obscured by the presence in the effluent of" "bacteria which have not come down through the filter directly" "from the Applied Water. Some of them appear to have their" "origin in the outlet-pipes and underdrains where they continue" "to live upon the very slight amount of food present. This is" "especially noticeable during the warm summer months when" "a few of the more hardy species grow upon the organic matter" "stored at the surface."

The principal object of table No. 22 is to show the number of times that Percentages of One Per cent. and More, of the Applied Water Bacteria, Appeared in the Filtered Water of the different filters, the number of times that the Percentages of Applied Water Bacteria Removed were One Per cent. and More Less than the Average Per cent Removed, and the Percentages that the number of times are of the Total Number of Results obtained; and to show the Percentages that the number of times that More than Two Per cent. of the Applied Water Bacteria Appeared in the Filtered Water, are of the Total Number of Results obtained, and the Percentages that the number of times, that the Percentages of the Applied Water Bacteria Removed that were More than Two Per cent. Less than the Average Per cent. Removed, are of the Total Number of Results obtained. The above results were calculated from data given in the tables of the Massachusetts Report, above mentioned, on pages from 491 to 524.

The average rates of filtration given in the Seventh column, of the First part of table No. 22 were obtained by averaging the daily rates of filtration, from June to November inclusive, of the days when both samples of Applied and Filtered Water were collected (given in the Massachusetts Report on pages from 491 to 524), with the exception of the rates of those days when the number of bacteria in the "Effluent" exceeded the number in the Applied Water, and are therefore somewhat approximate.

The Average Percentages of Bacteria Removed, not inclosed in parentheses, given in the Ninth column of the First part of table No. 22, were obtained by averaging the individual per cents worked out from daily samples, taken from June to November inclusive, given in the tables of the Massachusetts Report on pages from 491 to 524, with the exception of those cases in which the number of bacteria in the "Effluent" exceeded the number in the Applied Water, (viz: 2 in 33A; 3 in 34A; 3 in 36A; 2 in 37; 1 in 38; 1 in 39; 2 in 40). The Massachusetts Report states, in a note under the tables of December bacterial results, that "Channels were formed in the sides of the filters," and on page 477 that "This took place to a greater or less extent in the case" "of all the small filters, and the results obtained in December" "have not for this reason been included in the discussion." The number of results used in working out these percentages is given in the Second, Third and Fourth parts of table No. 22. The percentages were computed in the manner above described in order to compare them with the results obtained with the Morison Mechanical Filter, as, has previously been explained, the percentages given in the tables relating to the Morison Mechanical Filter, that have been considered in all comparisons and summaries, were obtained by using individual per cents which were worked out from all the results obtained while the filter was in its normal condition, there not being any results rejected on account of excessive numbers of bacteria being found in the Filtered Water.

The figures inclosed in parentheses, given in the Ninth column of the First part of table No. 22, were obtained by averaging the individual per cents of daily samples in the same manner as the percentages which are not inclosed in parentheses, with the exception that samples were not considered in which the number of bacteria in the "Effluent" exceeded 500. These percentages were computed by this method in order to show the difference that the rejection of the last mentioned samples would make in the average percentages.

A foot-note at the bottom of the tables in the Massachusetts Report states that "Numbers above 500 do not appear in the" "averages (see page 530)." The information on page 530 of the Massachusetts Report, referred to in this note, relating to the subject, is as follows: "The statistics in the tables (pp. 490–525)" "show that all of the effluents at times contained very large" "numbers of bacteria during July and August. In some cases" "they equalled and even exceeded the number applied. This" "was least noticeable in case of the intermittent filters Nos. 35 A" "and 41. Some error in the process of determination was at first" "suggested as the reason for this. Detailed study of the condi-" "tions under which the examinations were made, however, to-" "gether with the results of more numerous examinations, indi-" "cated that this was not so. It then appeared that there must" "be present in the filters at times conditions which favored the" "growth of certain kinds of bacteria."

The figures given in the Tenth column of the First part of table No. 22, not inclosed in parentheses, are percentages of removal worked out by using the total number of bacteria found in the "Effluent" and Applied Water of each filter during the entire period above specified, instead of from individual results. In other respects the same method was followed as was used in working out the percentages given in the Ninth column. These percentages were computed in order to compare them with the corresponding percentages given in the Ninth column.

The average percentages given in the Tenth column inclosed in parentheses, were worked out in the same manner as those not inclosed in parentheses in the Tenth column, with the exception that samples were not considered in which the number of bacteria in the "Effluent" exceeded 500. These percentages, which were also computed in order to compare them with the corresponding percentages given in the Ninth column, were worked out in a manner similar to the method followed in working out the bacterial percentages given in the Massachusetts Report.

The Average Percentages given in table No. 22, considered in making all comparisons with the results obtained with the Morison Mechanical Filter, unless otherwise specified, were computed by averaging the individual per cents of daily samples including samples in which the bacteria in the "Effluent" exceeded 500, with the exception of those samples in which the number in the "Effluent" exceeded the number in the Applied Water. As can be seen by the table the average per cents obtained by this method range from 0.5 to 2.8 less than the averages computed by using, as mentioned above, the total number of bacteria found in the "Effluent" and Applied Water.

TABLE NO. 2.

FILTRATION EXPERIMENTS.—MORISON MECHANICAL FILTER.

END GROWTHS, of Water Bacteria in the Samples of Applied and Filtered Water that were taken at the SAME HOUR (which was One Hour or More after water commenced to flow from the filter).

DATE.	Gallons of Water Filtered per Acre, per 24 Hours.	Bacteria per Cubic Centimeter.		Per cent. of the Applied Bacteria Removed.	Average Percentage of the Applied Bacteria Removed.	Grains of Sulphate of Alumina used per Gallon.
		In Applied Water.	In Filtered Water.			
1893.						
July * 20,	125,000,000	2,000	11	99.5	......	0.75
" 21,	122,000,000	9,477	16	99.8	......	0.90
Oct. 3,	125,000,000	905	6	99.3	......	0.60
" 4,	128,000,000	610	2	99.7	......	0.58
" 5,	131,000,000	4,002	25	99.4	99.5 (99.6)	0.55
Oct. 17,	125,000,000	6,175+	43	99.3	......	†0.57
" 27,	122,000,000	10,700	44	99.6	.....	0.61
" 30,	128,000,000	1,700	12	99.3	......	0.56
" 31,	131,000,000	500	16	96.8	......	0.59
Nov. 1,	132,000,000	21,200	28	99.9	......	0.61
" 2,	123,000,000	7,600	34	99.6	......	0.81
" 3,	122,000,000	12,500	66	99.5	......	0.84
" 4,	132,000,000	4,100	101	97.5	......	‡1.20
" 9,	125,000,000	3,300	35	98.9	......	0.85
" 11,	125,000,000	3,800	26	99.3	99.0 (99.4)	0.82
Commenced to use Bacillus Prodigiosus.						
Nov. 23,	120,000,000	15,850	218	98.6	......	0.60
" 24,	132,000,000	14,000	364	97.4	......	0.59
Dec. 2,	125,000,000	6,000	190	96.8	......	‖0.50
" 4,	128,000,000	4,475	91	98.0	......	0.60
1894.						
Jan. 2,	132,000,000	2,850	178	93.8	......	0.85
" 3,	137,000,000	3,375	192	94.3	......	0.84
" 4,	132,000,000	5,025	136	97.3	......	0.85
" 5,	130,000,000	3,775	142	96.2	......	0.82
" §8,	130,000,000	4,000	360	91.0	95.9 (96.8)	0.58
Ceased to use Bacillus Prodigiosus.						

TABLE NO. 2.—CONCLUDED.

DATE.		Gallons of Water Filtered per Acre, per 24 Hours.	Bacteria per Cubic Centimeter.		Per cent. of the Applied Bacteria Removed.	Average Percentage of the Applied Bacteria Removed.	Grains of Sulphate of Alumina used per Gallon.
			In Applied Water.	In Filtered Water.			
1894.							
Jan.	9,	130,000,000	3,400	148	95.6	......	0.60
"	10,	134,000,000	1,725	108	93.7	......	0.84
"	11,	130,000,000	2,150	84	96.1	......	0.61
"	12,	132,000,000	875	274	68.7	......	0.81
"	13,	132,000,000	1,633	68	95.8	......	0.72
"	15,	134,000,000	1,600	184	88.5	......	0.84
"	16,	134,000,000	775	178	77.0	......	0.58
"	17,	130,000,000	3,375	150	95.6	......	0.82
"	18,	134,000,000	3,800	162	95.7	......	0.59
"	19,	136,000,000	2,767	206	92.6	......	0.83
"	20,	130,000,000	5,200	230	95.6	......	0.72
"	22,	132,000,000	11,200	346	96.9	......	0.85
"	23,	132,000,000	4,133	278	93.3	91.2 (94.3)	0.80
Washed filter-bed with Caustic Soda.							
Jan.	24,	128,000,000	5,025	24	99.5	......	0.60
"	25,	125,000,000	3,600	36	99.0	......	0.82
"	26,	128,000,000	10,700	117	98.9	......	0.58
"	27,	128,000,000	7,567	123	98.4	......	0.58
"	29,	128,000,000	3,100	120	96.1	......	0.59
"	30,	130,000,000	3,000	70	97.7	98.3 (98.5)	0.58

*The counts from July 20 to July 21, and from Oct. 3 to Oct. 5, were made by Professor H. C. Ernst.

†One-half "Free Flow."

‡One-half grain when sample was taken.

Does not include "Free Flow" although it was used.

§Temperature of Applied Water 71°.

TABLE NO. 3.

FILTRATION EXPERIMENTS.—MORISON MECHANICAL FILTER.

Growths of about NINETY HOURS, of Water Bacteria in the Samples of Applied and Filtered Water that were taken at the SAME HOUR (which was One Hour or More after water commenced to flow from the filter).

DATE.	Gallons of Water Filtered per Acre, per 24 Hours.	Bacteria per Cubic Centimeter.		Per cent. of the Applied Bacteria Removed.	Average Percentage of the Applied Bacteria Removed.	Grains of Sulphate of Alumina used per Gallon.
		In Applied Water.	In Filtered Water.			
1893.						
July *20,	125,000,000	2,000	11	99.5	......	0.75
" 21,	122,000,000	9,477	16	99.8	......	0.90
Oct. 3,	125,000,000	905	6	99.3	......	0.60
" 4,	128,000,000	610	2	99.7	......	0.58
" 5,	131,000,000	4,002	25	99.4	99.5 (99.6)	0.55
Oct. 17,	125,000,000	6,175+	26	99.6	......	†0.57
" 27,	122,000,000	9,700	41	99.6	.....	0.61
" 30,	128,000,000	1,700	7	99.6	......	0.56
" 31,	131,000,000	400	9	97.8	......	0.59
Nov. 1,	132,000,000	15,112	19	99.9	......	0.61
" 2,	123,000,000	6,950	26	99.6	......	0.81
" 3,	122,000,000	9,400	50	99.5	......	0.84
" 4.	132,000,000	3,400	63	98.1	......	‡1.20
" 9,	125,000,000	2,200	26	98.8	......	0.60
" 11,	125,000,000	3,650	25	99.3	99.2 (99.5)	0.82
Commenced to use Bacillus Prodigiosus.						
Nov. 23,	120,000,000	15,850	218	98.6	......	0.60
" 24,	132,000,000	7,600	364	95.2	......	0.59
Dec. 2,	128,000,000	4,900	190	96.1	......	‖0.75
" 4,	128,000,000	4,475	91	98.0	......	0.60
1894.						
Jan. 2,	132,000,000	2,150	94	95.6	......	0.85
" 3,	137,000,000	2,000	118	94.1	......	0.84
" 4,	134,000,000	2,275	44	98.1	......	0.85
" 5,	130,000,000	1,925	60	96.9	......	0.82
" §8,	130,000,000	2,375	184	92.3	96.1 (96.9)	0.58
Ceased to use Bacillus Prodigiosus.						

TABLE NO. 3.—CONCLUDED.

DATE.		Gallons of Water Filtered per Acre, per 24 Hours.	Bacteria per Cubic Centimeter.		Per cent. of the Applied Bacteria Removed.	Average Percentage of the Applied Bacteria Removed.	Grains of Sulphate of Alumina used per Gallon.
			In Applied Water.	In Filtered Water.			
1894.							
Jan.	9,	130,000,000	1,850	54	97.1	......	0.60
"	10,	134,000,000	800	28	96.5	......	0.84
"	11,	130,000,000	750	20	97.3	......	0.61
"	12,	132,000,000	350	52	85.1	......	0.81
"	13,	132,000,000	600	36	94.0	.....	0.72
"	15,	134,000,000	925	88	90.5	......	0.84
"	16,	134,000,000	375	44	88.3	......	0.58
"	17,	130,000,000	2,150	64	97.0	......	0.82
"	18,	134,000,000	1,500	62	95.9	......	0.59
"	19,	136,000,000	1,450	80	94.5	......	0.83
"	20,	130,000,000	2,800	58	97.9	......	0.72
"	22,	132,000,000	3,350	62	98.1	......	0.85
"	23,	132,000,000	2,300	64	97.2	94.6 (96.3)	0.80
Washed filter-bed with Caustic Soda.							
Jan.	24,	128,000,000	2,100	6	99.7	......	0.60
"	25,	125,000,000	2,225	18	99.2	......	0.82
"	26,	128,000,000	4,650	54	98.8	......	0.58
"	27,	128,000,000	4,875	72	98.5	.....	0.58
"	29,	128,000,000	1,575	82	94.8	......	0.59
"	30,	130,000,000	1,400	28	98.0	98.2 (98.5)	0.58

* The counts from July 20 to July 21, and from Oct. 3 to Oct. 5, were made by Professor H. C. Ernst.

† One-half "Free Flow."

‡ One-half grain when sample was taken.

‖ Does not include "Free Flow" although it was used.

§ Temperature of Applied Water 71°.

TABLE NO. 4.

FILTRATION EXPERIMENTS.—MORISON MECHANICAL FILTER.

***END GROWTHS**, of Water Bacteria in all of the Samples of Filtered Water that were taken **ONE HOUR OR MORE** after water commenced to flow from the filter.*

Date.	Hour that Sample was taken.	Gallons of Water Filtered per Acre, per 24 Hours.	Bacteria per Cubic Centimeter.		Per cent. of the Applied Bacteria Removed.	Average Percentage of the Applied Bacteria Removed.	Grains of Sulphate of Alumina used per Gallon.
			In Applied Water.	In Filtered Water.			
July *20, 1893.	10.00 A.M.	125,000,000	2,000	11	99.5	......	0.75
" 21, "	12 M.	122,000,000	9,477	16	99.8	......	0.90
Oct. 3, "	12 M.	125,000,000	905	6	99.3	......	0.60
" 4, "	12 M.	128,000,000	610	2	99.7	......	0.58
" 5, "	12 M.	131,000,000	4,002	25	99.4	99.5	0.55
						(99.6)	
Oct. 17, "	12 M.	125,000,000	6,175+	43	99.3	......	† 0.57
" 24, "	3.00 P.M.	125,000,000	13,020+	529	95.9	......	0.60
" 27, "	12 M.	122,000,000	10,700	44	99.6	......	0.61
" 30, "	12 M.	128,000,000	1,700	12	99.3	......	0.56
" 31, "	12 M.	131,000,000	500	16	96.8	......	0.59
Nov. 1, "	12 M.	132,000,000	21,200	28	99.9	......	0.61
" 2, "	12 M.	123,000,000	7,600	34	99.6	......	0.81
" 3, "	12 M.	122,000,000	12,500	66	99.5	......	0.84
" 4, "	12 M.	132,000,000	4,100	101	97.5	......	‡ 1.20
" 9, "	10.15 A.M.	125,000,000	3,300	35	98.9	......	0.85
" 10, "	10.00 A.M. End.	128,000,000	4,750	96	98.0	......	0.85

TABLE NO. 4.—CONTINUED.

DATE.	Hour that Sample was taken.	Gallons of Water Filtered per Acre. per 24 Hours.	Bacteria per Cubic Centimeter.		Per cent. of the Applied Bacteria Removed.	Average Percentage of the Applied Bacteria Removed.	Grains of Sulphate of Alumina used per Gallon.
			In Applied Water.	In Filtered Water.			
Nov. 11, 1893.	3.35 P.M.	125,000,000	3,800	26	99.3	98.6 (98.8)	0.82
			Commenced to use Bacillus Prodigiosus.				
Nov. 23, 1893.	9.00 A.M.	116,000,000	15,850	202	98.7	......	0.60
	11.00 A.M.	120,000,000	15,850	161	99.0	......	0.60
	12 M.	120,000,000	15,850	218	98.6	......	0.60
	1.00 P.M.	125,000,000	15,850	223	98.6	......	0.60
	3.40 P.M.	132,000,000	15,850	272	98.3	......	0.60
" 24, "	5.43 A.M. End.	128,000,000	15,850	712	95.5	......	0.60
	11.00 A.M.	132,000,000	14,000	661	95.3	......	0.59
	12 M.	132,000,000	14,000	364	97.4	.....	0.59
	1.00 P.M.	128,000,000	14,000	371	97.4	......	0.59
	2.00 P.M.	121,000,000	14,000	140	99.0	......	0.59
Dec. 2, "	9.30 A.M.	128,000,000	6,000	132	97.8	......	‖ 0.50
	10.30 A.M.	125,000,000	6,000	220	96.3	......	‖ 0.50
	11.30 A.M.	125,000,000	6,000	190	96.8	......	‖ 0.50
	2.30 P.M.	125,000,000	6,000	84	98.6	......	‖ 0.75
	3.40 P.M.	125,000,000	6,000	120	98.0	......	‖ 0.75
" 4, "	12 M.	128,000,000	4,475	91	98.0	......	0.60
" 7, "	4.55 A.M.	128,000,000	4,000	80	98.0	......	0.59
" 8, "	11.00 A.M. End.	134,000,000	2,700	37	98.6	......	0.60
" 12, "	10.30 A.M.	125,000,000	3,200	132	95.9	......	0.58

Dec. 14, 1893.	9.30 A.M.		136,000,000	4,950	168	96.6	......	0.59
Jan. 2, 1894.	12.30 P.M.		132,000,000	2,850	178	93.8	......	0.85
	3.40 P.M.		130,000,000	2,850	190	93.3	......	0.85
" 3, "	12.30 P.M.		137,000,000	3,375	192	94.3	......	0.84
	1.30 P.M.		137,000,000	3,375	178	94.7	......	0.84
	3.40 P.M.		136,000,000	3,375	194	94.3	......	0.84
" 4, "	3.30 A.M.	End.	128,000,000	3,375	158	95.3	......	0.84
	1.30 P.M.		132,000,000	5,025	136	97.3	......	0.85
	3.40 P.M.		130,000,000	5,025	131	97.4	......	0.85
" 5, "	3.50 A.M.	End.	130,000,000	5,025	148	97.1	......	0.85
	12.30 P.M.		130,000,000	3,775	142	96.2	......	0.82
" 6, "	3.40 A.M.	End.	125,000,000	3,775	137	96.4	......	0.82
" §8, "	9.30 A.M.		142,000,000	4,000	402	90.0	......	0.58
	12.30 P.M.		130,000,000	4,000	360	91.0	......	0.58
	1.30 P.M.		128,000,000	4,000	372	90.7	......	0.58
	3.40 P.M.		128,000,000	4,000	480	88.0	96.1 (96.9)	0.58
			Ceased to use Bacillus Prodigiosus.					
Jan. 9, 1894.	9.30 A.M.		142,000,000	3,400	138	95.9	......	0.60
	12.30 P.M.		130,000,000	3,400	148	95.6	......	0.60
	1.30 P.M.		132,000,000	3,400	102	97.0	......	0.60
	3.40 P.M.		130,000,000	3,400	74	97.8	......	0.60
" 10, "	3.30 A.M.	End.	125,000,000	3,400	38	98.9	......	0.60
	9.30 A.M.		142,000,000	1,725	88	94.9	......	0.84
	12.30 P.M.		134,000,000	1,725	108	93.7	......	0.84
" 11, "	3.40 A.M.	End.	128,000,000	1,725	42	97.6	......	0.84
	9.30 A.M.		132,000,000	2,150	54	97.5	......	0.61
	12.30 P.M.		130,000,000	2,150	84	96.1	......	0.61
	1.30 P.M.		132,000,000	2,150	68	96.8	......	0.61

TABLE NO. 4.—CONTINUED.

Date.	Hour that Sample was taken.	Gallons of Water Filtered per Acre, per 24 Hours.	Bacteria per Cubic Centimeter.		Per cent. of the Applied Bacteria Removed.	Average Percentage of the Applied Bacteria Removed.	Grains of Sulphate of Alumina used per Gallon.
			In Applied Water.	In Filtered Water.			
Jan. 11, 1894.	3.40 P.M.	128,000,000	2,150	74	96.6	......	0.61
" 12, "	5.30 A.M. End.	125,000,000	2,150	66	96.9	......	0.61
	9.30 A.M.	134,000,000	875	132	84.9	......	0.81
	12.30 P.M.	132,000,000	875	274	68.7	......	0.81
	1.30 P.M.	130,000,000	875	90	89.7	......	0.81
	3.40 P.M.	128,000,000	875	98	88.8	......	0.81
" 13, "	3.40 A.M. End.	121,000,000	875	38	95.7	......	0.81
	9.30 A.M.	130,000,000	1,633	72	95.6	......	0.72
	12.30 P.M.	132,000,000	1,633	68	95.8	......	0.72
	1.30 P.M.	130,000,000	1,633	66	96.0	......	0.72
	3.30 P.M.	128,000,000	1,633	48	97.1	......	0.72
" 15, "	9.30 A.M.	132,000,000	1,600	104	93.5	......	0.84
	12.30 P.M.	134,000,000	1,600	184	88.5	......	0.84
	1.30 P.M.	136,000,000	1,600	294	81.6	......	0.84
	3.40 P.M.	132,000,000	1,600	318	80.1	......	0.84
" 16, "	3.50 A.M. End.	125,000,000	1,600	264	83.5	......	0.84
	9.30 A.M.	142,000,000	775	162	79.1	......	0.58
	12.30 P.M.	134,000,000	775	178	77.0	......	0.58
	1.30 P.M.	134,000,000	775	142	81.7	......	0.58
	3.40 P.M.	132,000,000	775	150	80.6	......	0.58
" 17, "	5.30 A.M. End.	128,000,000	775	134	82.7	......	0.58
	9.30 A.M.	128,000,000	3,375	162	95.2	......	0.82
	12.30 P.M.	130,000,000	3,375	150	95.6	......	0.82

Jan. 17, 1894.	1.30 P.M.		130,000,000	3,375	142	95.8	......	0.82
	3.40 P.M.		128,000,000	3,375	150	95.6	......	0.82
" 18, "	3.40 A.M.	End.	125,000,000	3,375	74	97.8	......	0.82
	9.30 A.M.		132,000,000	3,800	142	96.3	......	0.59
	12.30 P.M.		134,000,000	3,800	162	95.7	......	0.59
	1.30 P.M.		132,000,000	3,800	186	95.1	......	0.59
	3.40 P.M.		130,000,000	3,800	154	95.9	......	0.59
" 19, "	5.30 A.M.	End.	128,000,000	3,800	82	97.8	......	0.59
	9.30 A.M.		142,000,000	2,767	204	92.6	......	0.83
	12.30 P.M.		136,000,000	2,767	206	92.6	......	0.83
	1.30 P.M.		134,000,000	2,767	160	94.2	......	0.83
	3.40 P.M.		132,000,000	2,767	148	94.7	......	0.83
" 20, "	3.50 A.M.	End.	128,000,000	2,767	116	95.8	......	0.83
	9.30 A.M.		136,000,000	5,200	150	97.1	......	0.72
	12.30 P.M.		130,000,000	5,200	230	95.6	......	0.72
	1.30 P.M.		130,000,000	5,200	234	95.5	......	0.72
	3.30 P.M.		130,000,000	5,200	216	95.8	......	0.72
" 22, "	9.30 A.M.		142,000,000	11,200	396	96.5	......	0.85
	12.30 P.M.		132,000,000	11,200	346	96.9	......	0.85
	1.30 P.M.		132,000,000	11,200	382	96.6	......	0.85
	3.40 P.M.		130,000,000	11,200	344	96.9	......	0.85
" 23, "	3.45 A.M.	End.	125,000,000	11,200	222	98.0	......	0.85
	9.30 A.M.		142,000,000	4,133	258	93.8	......	0.80
	12.30 P.M.		132,000,000	4,133	278	93.3	92.8 (95.2)	0.80
			Washed filter-bed with Caustic Soda.					
Jan. 24, 1894.	9.30 A.M.		132,000,000	5,025	42	99.2	......	0.60
	12.30 P.M.		128,000,000	5,025	24	99.5	......	0.60
	1.30 P.M.		128,000,000	5,025	30	99.4	......	0.60
	3.40 P.M.		128,000,000	5,025	34	99.3	......	0.60

TABLE NO. 4.—CONCLUDED.

DATE.	Hour that Sample was taken.		Gallons of Water Filtered per Acre. per 24 Hours.	Bacteria per Cubic Centimeter.		Per cent. of the Applied Bacteria Removed.	Average Percentage of the Applied Bacteria Removed.	Grains of Sulphate of Alumina used per Gallon.
				In Applied Water.	In Filtered Water.			
Jan. 25, 1894.	5.30 A.M.	End.	128,000,000	5,025	32	99.4	......	0.60
	9.30 A.M.		128,000,000	3,600	55	98.5	......	0.82
	12.30 P.M.		125,000,000	3,600	36	99.0	......	0.82
	1.30 P.M.		125,000,000	3,600	30	99.2	......	0.82
	3.40 P.M.		128,000,000	3,600	40	98.9	......	0.82
" 26, "	3.45 A.M.		121,000,000	3,600	34	99.1	......	0.82
	12 M.		128,000,000	10,700	117	98.9	......	0.58
	3.40 P.M.		128,000,000	10,700	66	99.4	......	0.58
" 27, "	5.35 A.M.	End.	125,000,000	10,700	86	99.2	......	0.58
	9.30 A.M.		128,000,000	7,567	135	98.2	......	0.58
	12.30 P.M.		128,000,000	7,567	123	98.4	......	0.58
" 29, "	12 M.		128,000,000	3,100	120	96.1	......	0.59
" 30, "	5.30 A.M.	End.	128,000,000	3,100	54	98.3	......	0.59
	12 M.		130,000,000	3,000	70	97.7	98.8 (98.9)	0.58

* The counts from July 20 to July 21, and from Oct. 3 to Oct. 5, were made by Prof. H. C. Ernst.

† One-half "Free Flow."

‡ One-half grain when sample was taken.

[1] Does not include "Free Flow" although it was used.

§ Temperature of Applied Water on this date 71°.

"End" in second column signifies end of run.

Table No. 5.

Filtration Experiments.—Morison Mechanical Filter.

Growths of Water Bacteria of EIGHTY FIVE HOURS OR MORE AND END GROWTHS, *in all of the Samples of Filtered Water that were taken* ONE HOUR OR MORE *after water commenced to flow from the filter.*

Date.	Hour that Sample was taken.	Gallons of Water Filtered per Acre, per 24 Hours.	Bacteria per Cubic Centimeter.		Per cent. of the Applied Bacteria Removed.	Average Percentage of the Applied Bacteria Removed.	Grains of Sulphate of Alumina used per Gallon.
			In Applied Water.	In Filtered Water.			
Oct. 17, 1893.	12 M.	125,000,000	6,175+	43	99.3	......	*0.57
" 23, "	12 M.	196,000,000	11,700	76	99.4	......	0.86
" 24, "	3.00 P.M.	125,000,000	13,020+	529	95.9	......	0.60
" 27, "	12 M.	122,000,000	10,700	44	99.6	......	0.61
" 30, "	12 M.	128,000,000	1,700	12	99.3	......	0.56
" 31, "	12 M.	131,000,000	500	16	96.8	......	0.59
Nov. 1, "	12 M.	132,000,000	21,200	28	99.9	.....	0.61
" 2, "	12 M.	123,000,000	7,600	34	99.6	......	0.81
" 3, "	12 M.	122,000,000	12,500	66	99.5	.....	0.84
" 4, "	12 M.	132,000,000	4,100	101	97.5	......	†1.20
" 9, "	10.15 A.M.	125,000,000	3,300	35	98.9	......	0.85
" 10, "	10.00 A.M.	128,000,000	4,750	96	98.0	......	0.85
" 11, "	3.35 P.M.	125,000,000	3,800	26	99.3	.....	0.82
" 15, "	3.35 P.M.	125,000,000	3,800	52	98.6	......	0.88
" 17, "	3.35 P.M.	125,000,000	24,600	70	99.7	.....	0.85

TABLE NO. 5.—CONTINUED.

DATE.	Hour that Sample was taken.	Gallons of Water Filtered per Acre, per 24 Hours.	Bacteria per Cubic Centimeter.		Per cent. of the Applied Bacteria Removed.	Average Percentage of the Applied Bacteria Removed.	Grains of Sulphate of Alumina used per Gallon.
			In Applied Water.	In Filtered Water.			
Nov. 18, 1893.	3.35 P.M.	132,000,000	15,700	55	99.6	98.8 (99.1)	0.61
			Commenced to use Bacillus Prodigiosus.				
Nov. 23, 1893.	9.00 A.M.	116,000,000	15,850	202	98.7	......	0.60
	11.00 A.M.	120,000,000	15,850	161	99.0	......	0.60
	12 M.	120,000,000	15,850	218	98.6	......	0.60
	1.00 P.M.	125,000,000	15,850	223	98.6	.....	0.60
	2.00 P.M.	128,000,000	15,850	198	98.8	......	0.60
	3.00 P.M.	132,000,000	15,850	193	98.8	......	0.60
	3.40 P.M.	132,000,000	15,850	272	98.3	......	0.60
" 24, "	5.43 A.M.	128,000,000	15,850	712	95.5	......	0.60
	9.00 A.M.	146,000,000	14,000	705	95.0	......	0.59
	10.00 A.M.	146,000,000	14,000	441	96.9	......	0.59
	11.00 A.M.	132,000,000	14,000	661	95.3	......	0.59
	12 M.	132,000,000	14,000	364	97.4	......	0.59
	1.00 P.M.	128,000,000	14,000	371	97.4	......	0.59
	2.00 P.M.	121,000,000	14,000	140	99.0	......	0.59
	3.00 P.M.	121,000,000	14,000	130	99.1	......	0.59
	3.40 P.M.	132,000,000	14,000	137	99.0	......	0.59
" 25, "	10.00 A.M.	125,000,000	6,400	294	95.4	......	0.61
" 28, "	9.30 A.M.	128,000,000	4,200	336	92.0	......	0.61
	10.30 A.M.	132,000,000	4,200	210	95.0	......	0.61

Nov. 28, 1893.	11.30 A.M.	132,000,000	4,200	234	94.4	......	0.61
	12.15 P.M.	128,000,000	4,200	57	98.6	......	0.61
	12.30 P.M.	128,000,000	4,200	224	94.7	......	0.61
Dec. 2, "	9.30 A.M.	128,000,000	6,000	132	97.8	......	‡0.50
	10.30 A.M.	125,000,000	6,000	220	96.3	......	‡0.50
	11.30 A.M.	125,000,000	6,000	190	96.8	......	‡0.50
	12.30 P.M.	128,000,000	6,000	143	97.6	......	‡0.75
	1.30 P.M.	128,000,000	6,000	103	98.3	......	‡0.75
	2.30 P.M.	125,000,000	6,000	84	98.6	......	‡0.75
	3.40 P.M.	125,000,000	6,000	120	98.0	......	‡0.75
" 4, "	12 M.	128,000,000	4,475	91	98.0	......	0.60
" 7, "	4.55 A.M.	128,000,000	4,000	80	98.0	......	0.59
" 8, "	11.00 A.M.	134,000,000	2,700	37	98.6	......	0.60
" 9, "	10.00 A.M.	132,000,000	3,400	41	98.8	......	0.58
" 12, "	9.30 A.M.	128,000,000	3,200	130	95.9	......	0.58
	10.30 A.M.	125,000,000	3,200	132	95.9	......	0.58
	11.30 A.M.	128,000,000	3,200	220	93.1	......	0.58
	12.30 P.M.	128,000,000	3,200	133	95.8	......	0.58
	1.30 P.M.	125,000,000	3,200	69	97.8	......	0.58
	2.30 P.M.	131,000,000	3,200	118	96.3	......	0.58
	3.40 P.M.	128,000,000	3,200	101	96.8	......	0.58
" 13, "	5.30 A.M.	128,000,000	3,200	65	98.0	......	0.58
	9.30 A.M.	134,000,000	3,833	217	94.3	......	0.60
	10.30 A.M.	134,000,000	3,833	194	94.9	......	0.60
	11.30 A.M.	137,000,000	3,833	156	95.9	......	0.60
	12.30 P.M.	134,000,000	3,833	173	95.5	......	0.60
	1.30 P.M.	134,000,000	3,833	150	96.1	......	0.60
	2.30 P.M.	130,000,000	3,833	121	96.8	......	0.60
	3.40 P.M.	132,000,000	3,833	124	96.8	......	0.60
" 14, "	5.15 A.M.	132,000,000	3,833	71	98.1	......	0.60

TABLE NO. 5.—CONTINUED.

DATE.	Hour that Sample was taken.	Gallons of Water Filtered per Acre, per 24 Hours.	Bacteria per Cubic Centimeter.		Per cent. of the Applied Bacteria Removed.	Average Percentage of the Applied Bacteria Removed.	Grains of Sulphate of Alumina used per Gallon.
			In Applied Water.	In Filtered Water.			
Dec. 14, 1893.	9.30 A.M.	136,000,000	4,950	168	96.6	......	0.59
	10.30 A.M.	137,000,000	4,950	201	95.9	......	0.59
	11.30 A.M.	136,000,000	4,950	73	98.5	......	0.59
	12.30 P.M.	134,000,000	4,950	172	96.5	......	0.59
	1.30 P.M.	134,000,000	4,950	73	98.5	......	0.59
	2.30 P.M.	134,000,000	4,950	88	98.2	......	0.59
	3.40 P.M.	132,000,000	4,950	132	97.3	......	0.59
" 15, "	5.35 A.M.	132,000,000	1,900	26	98.6	......	0.59
	9.30 A.M.	137,000,000	1,900	57	97.0	......	0.66
	10.30 A.M.	132,000,000	1,900	83	95.6	......	0.66
	11.30 A.M.	132,000,000	1,900	59	96.9	......	0.66
	12.30 P.M.	134,000,000	1,900	71	96.3	......	0.66
	1.30 P.M.	130,000,000	1,900	61	96.8	......	0.66
	2.30 P.M.	128,000,000	1,900	53	97.2	......	0.66
	3.40 P.M.	132,000,000	1,900	83	95.6	......	0.66
" 16, "	4.35 A.M.	132,000,000	1,900	44	97.7	......	0.66
	10.00 A.M.	132,000,000	5,900	176	97.0	......	0.74
	11.00 A.M.	132,000,000	5,900	184	96.9	......	0.74
	12 M.	134,000,000	5,900	182	96.9	......	0.74
	1.00 P.M.	132,000,000	5,900	138	97.7	......	0.74
	2.00 P.M.	131,000,000	5,900	78	98.7	......	0.74
	3.00 P.M.	134,000,000	5,900	96	98.4	......	0.74
	3.30 P.M.	131,000,000	5,900	94	98.4	......	0.74

Dec. 18, 1893.	9.30 A.M.	131,000,000	8,200	228	97.2	......	0.59
	1.30 P.M.	128,000,000	8,200	336	95.9	......	0.59
	2.30 P.M.	131,000,000	8,200	322	96.1	......	0.59
	3.40 P.M.	128,000,000	8,200	302	96.3	......	0.59
" 19, "	4.40 A.M.	128,000,000	8,200	102	98.8	......	0.59
" 20, "	4.35 A.M.	132,000,000	5,350	124	97.7	......	0.58
	1.30 P.M.	128,000,000	5,350	112	97.9	......	0.60
	2.30 P.M.	130,000,000	5,350	126	97.6	......	0.60
	3.40 P.M.	128,000,000	5,350	116	97.8	......	0.60
" 21, "	5.30 A.M.	130,000,000	5,350	60	98.9	......	0.60
	9.30 A.M.	156,000,000	6,400	76	98.8	......	0.61
	10.30 A.M.	138,000,000	6,400	58	99.1	......	0.61
	11.30 A.M.	132,000,000	6,400	80	98.8	......	0.61
	12.30 P.M.	132,000,000	6,400	69	98.9	......	0.61
	1.30 P.M.	130,000,000	6,400	61	99.0	......	0.61
	2.30 P.M.	128,000,000	6,400	74	98.8	......	0.61
	3.40 P.M.	121,000,000	6,400	100	98.4	......	0.61
" 23, "	9.30 A.M.	146,000,000	6,100	136	97.8	......	0.72
	10.30 A.M.	142,000,000	6,100	152	97.5	......	0.72
	11.30 A.M.	136,000,000	6,100	151	97.5	......	0.72
	12.30 P.M.	134,000,000	6,100	95	98.4	......	0.72
	1.30 P.M.	132,000,000	6,100	119	98.0	......	0.72
	2.30 P.M.	132,000,000	6,100	121	98.0	......	0.72
	3.30 P.M.	132,000,000	6,100	206	96.6	......	0.72
Jan. 1, 1894.	9.30 A.M.	132,000,000	2,100	400	81.0	......	0.82
	1.30 P.M.	132,000,000	2,775	358	87.1	......	0.82
	3.40 P.M.	130,000,000	2,775	630	77.3	......	0.82
" 2, "	3.35 A.M.	128,000,000	2,525	770	69.5	......	0.82
	9.30 A.M.	147,000,000	2,775	158	94.3	......	0.85
	12.30 P.M.	132,000,000	2,850	178	93.8	......	0.85

TABLE NO. 5.—CONTINUED.

Date.	Hour that Sample was taken.	Gallons of Water Filtered per Acre, per 24 Hours.	Bacteria per Cubic Centimeter. In Applied Water.	Bacteria per Cubic Centimeter. In Filtered Water.	Per cent. of the Applied Bacteria Removed.	Average Percentage of the Applied Bacteria Removed.	Grains of Sulphate of Alumina used per Gallon.
Jan. 2, 1894.	1.30 P.M.	132,000,000	2,525	264	89.5	......	0.85
	3.40 P.M.	130,000,000	2,850	190	93.3	......	0.85
" 3, "	4.30 A.M.	130,000,000	2,850	136	95.2	......	0.85
	9.30 A.M.	142,000,000	2,725	64	97.7	......	0.84
	12.30 P.M.	137,000,000	3,375	192	94.3	......	0.84
	1.30 P.M.	137,000,000	3,375	178	94.7	......	0.84
	3.40 P.M.	136,000,000	3,375	194	94.3	......	0.84
" 4, "	3.30 A.M.	128,000,000	3,375	158	95.3	......	0.84
	9.30 A.M.	148,000,000	3,500	72	97.9	......	0.85
	12.30 P.M.	134,000,000	3,500	44	98.7	......	0.85
	1.30 P.M.	132,000,000	5,025	136	97.3	......	0.85
	3.40 P.M.	130,000,000	5,025	131	97.4	......	0.85
" 5, "	3.50 A.M.	130,000,000	5,025	148	97.1	......	0.85
	9.30 A.M.	136,000,000	3,600	154	95.7	......	0.82
	12.30 P.M.	130,000,000	3,775	142	96.2	......	0.82
	1.30 P.M.	132,000,000	3,600	148	95.9	......	0.82
	3.40 P.M.	130,000,000	3,775	382	89.9	......	0.82
" 6, "	3.40 A.M.	125,000,000	3,775	137	96.4	......	0.82
||	12.30 P.M.	146,000,000	2,700	332	87.7	......	1.00
||	1.30 P.M.	138,000,000	2,700	196	92.7	......	1.00
||	3.40 P.M.	132,000,000	2,700	186	93.1	......	1.00
" || 8, "	9.30 A.M.	142,000,000	4,000	402	90.0	......	0.58
||	12.30 P.M.	130,000,000	4,000	360	91.0	......	0.58

Jan. ‖ 8, 1894.	1.30 P.M.	128,000,000	4,000	372	90.7	......	0.58
‖	3.40 P.M.	128,000,000	4,000	480	88.0	......	0.58
" 9, "	5.30 A.M.	128,000,000	4,000	568	85.8	95.9 (96.8)	0.58
		Ceased to use Bacillus Prodigiosus.					
Jan. 9, 1894.	9.30 A.M.	142,000,000	3,400	138	95.9	......	0.60
	12.30 P.M.	130,000,000	3,400	148	95.6	......	0.60
	1.30 P.M.	132,000,000	3,400	102	97.0	......	0.60
	3.40 P.M.	130,000,000	3,400	74	97.8	......	0.60
" 10, "	3.30 A.M.	125,000,000	3,400	38	98.9	......	0.60
	9.30 A.M.	142,000,000	1,725	88	94.9	......	0.84
	12.30 P.M.	134,000,000	1,725	108	93.7	......	0.84
" 11, "	3.40 A.M.	128,000,000	1,725	42	97.6	......	0.84
	9.30 A.M.	132,000,000	2,150	54	97.5	......	0.61
	12.30 P.M.	130,000,000	2,150	84	96.1	......	0.61
	1.30 P.M.	132,000,000	2,150	68	96.8	......	0.61
	3.40 P.M.	128,000,000	2,150	74	96.6	......	0.61
" 12, "	5.30 A.M.	125,000,000	2,150	66	96.9	......	0.61
	9.30 A.M.	134,000,000	875	132	84.9	......	0.81
	12.30 P.M.	132,000,000	875	274	68.7	......	0.81
	1.30 P.M.	130,000,000	875	90	89.7	......	0.81
	3.40 P.M.	128,000,000	875	98	88.8	......	0.81
" 13, "	3.40 A.M.	121,000,000	875	38	95.7	......	0.81
	9.30 A.M.	130,000,000	1,633	72	95.6	......	0.72
	12.30 P.M.	132,000,000	1,633	68	95.8	......	0.72
	1.30 P.M.	130,000,000	1,633	66	96.0	......	0.72
	3.30 P.M.	128,000,000	1,633	48	97.1	......	0.72
" 15, "	9.30 A.M.	132,000,000	1,600	104	93.5	......	0.84
	12.30 P.M.	134,000,000	1,600	184	88.5	......	0.84
	1.30 P.M.	136,000,000	1,600	294	81.6	......	0.84

TABLE NO. 5.—CONCLUDED.

Date.	Hour that Sample was taken.	Gallons of Water Filtered per Acre, per 24 Hours.	Bacteria per Cubic Centimeter.		Per cent. of the Applied Bacteria Removed.	Average Percentage of the Applied Bacteria Removed.	Grains of Sulphate of Alumina used per Gallon.
			In Applied Water.	In Filtered Water.			
Jan. 15, 1894.	3.40 P.M.	132,000,000	1,600	318	80.1	......	0.84
" 16, "	3.50 A.M.	125,000,000	1,600	264	83.5	......	0.84
	9.30 A.M.	142,000,000	775	162	79.1	......	0.58
	12.30 P.M.	134,000,000	775	178	77.0	......	0.58
	1.30 P.M.	134,000,000	775	142	81.7	......	0.58
	3.40 P.M.	132,000,000	775	150	80.6	......	0.58
" 17, "	5.30 A.M.	128,000,000	775	134	82.7	......	0.58
	9.30 A.M.	128,000,000	3,375	162	95.2	......	0.82
	12.30 P.M.	130,000,000	3,375	150	95.6	......	0.82
	1.30 P.M.	130,000,000	3,375	142	95.8	......	0.82
	3.40 P.M.	128,000,000	3,375	150	95.6	......	0.82
" 18, "	3.40 A.M.	125,000,000	3,375	74	97.8	......	0.82
	9.30 A.M.	132,000,000	3,800	142	96.3	......	0.59
	12.30 P.M.	134,000,000	3,800	162	95.7	......	0.59
	1.30 P.M.	132,000,000	3,800	186	95.1	......	0.59
	3.40 P.M.	130,000,000	3,800	154	95.9	......	0.59
" 19, "	5.30 A.M.	128,000,000	3,800	82	97.8	......	0.59
	9.30 A.M.	142,000,000	2,767	204	92.6	......	0.83
	12.30 P.M.	136,000,000	2,767	206	92.6	......	0.83
	1.30 P.M.	134,000,000	2,767	160	94.2	......	0.83
	3.40 P.M.	132,000,000	2,767	148	94.7	......	0.83
" 20, "	3.50 A.M.	128,000,000	2,767	116	95.8	......	0.83
	9.30 A.M.	136,000,000	5,200	150	97.1	......	0.72
	12.30 P.M.	130,000,000	5,200	230	95.6	......	0.72

Jan.	20, 1894.	1.30 P.M.	130,000,000	5,200	234	95.5	......	0.72
		3.30 P.M.	130,000,000	5,200	216	95.8	......	0.72
"	22, "	9.30 A.M.	142,000,000	11,200	396	96.5	......	0.85
		12.30 P.M.	132,000,000	11,200	346	96.9	......	0.85
		1.30 P.M.	132,000,000	11,200	382	96.6	......	0.85
		3.40 P.M.	130,000,000	11,200	344	96.9	......	0.85
"	23, "	3.45 A.M.	125,000,000	11,200	222	98.0	......	0.85
		9.30 A.M.	142,000,000	4,133	258	93.8	......	0.80
		12.30 P.M.	132,000,000	4,133	278	93.3	92.8 (95.2)	0.80
				Washed filter-bed with Caustic Soda.				
Jan.	24, 1894.	9.30 A.M.	132,000,000	5,025	42	99.2	......	0.60
		12.30 P.M.	128,000,000	5,025	24	99.5	......	0.60
		1.30 P.M.	128,000,000	5,025	30	99.4	......	0.60
		3.40 P.M.	128,000,000	5,025	34	99.3	......	0.60
"	25, "	5.30 A.M.	128,000,000	5,025	32	99.4	......	0.60
		9.30 A.M.	128,000,000	3,600	55	98.5	......	0.82
		12.30 P.M.	125,000,000	3,600	36	99.0	......	0.82
		1.30 P.M.	125,000,000	3,600	30	99.2	......	0.82
		3.40 P.M.	128,000,000	3,600	40	98.9	......	0.82
"	26, "	3.45 A.M.	121,000,000	3,600	34	99.1	......	0.82
		12 M.	128,000,000	10,700	117	98.9	......	0.58
		3.40 P.M.	128,000,000	10,700	66	99.4	......	0.58
"	27, "	5.35 A.M.	125,000,000	10,700	86	99.2	......	0.58
		9.30 A.M.	128,000,000	7,567	135	98.2	......	0.58
		12.30 P.M.	128,000,000	7,567	123	98.4	......	0.58
"	29, "	12 M.	128,000,000	3,100	120	96.1	......	0.59
"	30, "	5.30 A.M.	128,000,000	3,100	54	98.3	......	0.59
		12 M.	130,000,000	3,000	70	97.7	98.8 (98.9)	0.58

* One-half " Free Flow."
† One-half grain when sample was taken.
‡ Does not include " Free Flow " although it was used.
" Temperature of Applied Water 71°.

10

TABLE No. 6.

FILTRATION EXPERIMENTS.—MORISON MECHANICAL FILTER.

END GROWTHS, of Water Bacteria in the Samples of Filtered Water that were taken THIRTY MINUTES OR LESS after water commenced to flow from the filter.

Date.	Minutes after Flow.	Gallons of Water Filtered per Acre, per 24 Hours.	Bacteria per Cubic Centimeter.		Per cent. of the Applied Bacteria Removed.	Average Percentage of the Applied Bacteria Removed.	Grains of Sulphate of Alumina used per Gallon.
			In Applied Water.	In Filtered Water.			
Oct. * 6, 1893	20	131,000,000	7,795	60	99.2	99.2	0.60
Oct. 17, "	21	110,000,000	6,175+	82	98.7	......	† 0.57
" 27, "	18	116,000,000	10,700	431	96.0	......	0.61
" 30, "	18	106,000,000	1,700	66	96.1	......	0.56
Nov. 2, "	16	116,000,000	7,900	134	98.3	......	0.81
" 3, "	22	116,000,000	11,450	113	99.0	......	0.84
" 4, "	18	102,000,000	4,100	344	91.6	.....	‡ 1.20
" 11, "	15	104,000,000	3,650	252	93.1	96.1 (96.9)	0.82
			Commenced to use Bacillus Prodigiosus.				
Nov. 23, 1893	21	99,000,000	15,850	536	96.6	.....	0.60
" 24, "	30	132,000,000	14,000	1,103	92.1	......	0.59
Dec. 1, "	30	128,000,000	2,850	420	85.3	......	0.97
" 2, "	30	112,000,000	6,000	270	95.5	......	0.74
" 4, "	30	119,000,000	4,475	278	93.8	.....	0.60
" 7, "	30	98,000,000	4,000	91	97.7	......	0.60
Jan. 1, 1894	30	128,000,000	2,533	708	72.0	......	0.82
" 2, "	30	128,000,000	2,850	358	87.4	......	0.85

Jan. 3, 1894............	30	136,000,000	3,375	304	91.0	......	0.84
" 4, "	30	148,000,000	5,025	202	96.0	......	0.85
" 5, "	30	134,000,000	3,775	226	94.0	......	0.82
" ‖6, "	30	136,000,000	2,700	256	90.5	......	1.00
" ‖8, "	30	137,000,000	4,000	430	89.3	90.9 (92.7)	0.58
Ceased to use Bacillus Prodigiosus.							
Jan. 9, 1894............	30	132,000,000	3,400	156	95.4	......	0.60
" 10, "	30	128,000,000	1,725	132	92.3	......	0.84
" 11, "	30	121,000,000	2,150	100	95.3	......	0.61
" 12, "	30	134,000,000	875	94	89.3	......	0.81
" 13, "	30	128,000,000	1,633	106	93.5	......	0.72
" 15, "	30	134,000,000	1,600	116	92.8	......	0.84
" 16, "	30	134,000,000	775	296	61.8	......	0.58
" 17, "	30	113,000,000	3,375	350	89.6	......	0.82
" 18, "	30	121,000,000	3,800	234	93.8	......	0.59
" 19, "	30	132,000,000	2,767	192	93.1	......	0.83
" 20, "	30	125,000,000	5,200	208	96.0	......	0.72
" 22, "	30	134,000,000	11,200	682	93.9	......	0.85
" 23, "	30	128,000,000	4,133	514	87.6	90.3 (92.5)	0.80
Washed filter-bed with Caustic Soda.							
Jan. 24, 1894............	30	128,000,000	5,025	166	96.7	......	0.60
" 25, "	30	128,000,000	3,600	144	96.0	......	0.82
" 26, "	30	128,000,000	10,700	286	97.3	......	0.58
" 27, "	30	132,000,000	7,567	77	99.0	......	‡ 1.01
" 29, "	30	128,000,000	3,100	76	97.5	......	0.59
" 30, "	30	128,000,000	3,000	312	89.6	96.0 (96.8)	0.58

* The count on Oct. 6 was made by Professor H. C. Ernst.
† One-half "Free Flow."
‡ One-half grain when samples were taken.
Temperature of Applied Water 71°.

Table No. 7.

Filtration Experiments.—Morison Mechanical Filter.

Growths of about NINETY HOURS, of Water Bacteria in the Samples of Filtered Water that were taken THIRTY MINUTES OR LESS after water commenced to flow from the filter.

Date.	Minutes after Flow.	Gallons of Water Filtered per Acre, per 24 Hours.	Bacteria per Cubic Centimeter.		Per cent. of the Applied Bacteria Removed.	Average Percentage of the Applied Bacteria Removed.	Grains of Sulphate of Alumina used per Gallon.
			In Applied Water.	In Filtered Water.			
Oct. * 6, 1893	20	131,000,000	7,795	60	99.2	99.2	0.60
Oct. 17, "	21	110,000,000	6,175+	59	99.0	......	† 0.57
" 27, "	18	116,000,000	9,700	327	96.6	......	0.61
" 30, "	18	106,000,000	833	53	93.6	......	0.56
Nov. 2, "	16	116,000,000	6,950	92	98.7	......	0.81
" 3, "	22	116,000,000	9,400	82	99.1	......	0.84
" 4, "	18	102,000,000	3,400	184	94.6	......	‡ 1.20
" 11, "	15	104,000,000	3,650	252	93.1	96.4 (97.4)	0.82
			Commenced to use Bacillus Prodigiosus.				
Nov. 23, 1893	21	99,000,000	15,850	536	96.6	......	0.60
" 24, "	30	132,000,000	7,600	1,103	85.5	......	0.59
Dec. 1, "	30	128,000,000	2,500	420	83.2	......	0.97
" 2, "	30	112,000,000	4,900	270	94.5	......	0.74
" 4, "	30	119,000,000	4,475	278	93.8	......	0.60
" 7, "	30	98,000,000	4,000	91	97.7	......	0.60
Jan. 1, 1894	30	128,000,000	1,325	352	73.4	......	0.82
" 2, "	30	128,000,000	2,850	234	91.8	......	0.85

Jan. 3, 1894	30	136,000,000	2,000	142	92.9	......	0.84
" 4, "	30	148,000,000	2,275	76	96.7	......	0.85
" 5, "	30	134,000,000	1,925	92	95.2	......	0.82
" ‖ 6, "	30	136,000,000	1,400	158	88.7	......	1.00
" ‖ 8, "	30	137,000,000	2,375	264	88.9	90.7 (92.5)	0.58
		Ceased to use Bacillus Prodigiosus.					
Jan. 9, 1894	30	132,000,000	1,850	60	96.8	......	0.60
" 10, "	30	128,000,000	800	72	91.0	......	0.84
" 11, "	30	121,000,000	750	50	93.3	......	0.61
" 12, "	30	134,000,000	350	54	84.6	......	0.81
" 13, "	30	128,000,000	600	60	90.0	......	0.72
" 15, "	30	134,000,000	925	54	94.2	......	0.84
" 16, "	30	134,000,000	375	54	85.6	......	0.58
" 17, "	30	113,000,000	2,150	100	95.3	......	0.82
" 18, "	30	121,000,000	1,500	108	92.8	......	0.59
" 19, "	30	132,000,000	1,450	92	93.7	......	0.83
" 20, "	30	125,000,000	2,800	64	97.7	......	0.72
" 22, "	30	134,000,000	3,350	76	97.7	......	0.85
" 23, "	30	128,000,000	2,300	94	95.9	93.0 (95.1)	0.80
		Washed filter-bed with Caustic Soda.					
Jan. 24, 1894	30	128,000,000	2,100	88	95.8	......	0.60
" 25, "	30	128,000,000	2,225	108	95.1	......	0.82
" 26, "	30	128,000,000	4,650	136	97.1	......	0.58
" 27, "	30	132,000,000	4,875	40	99.2	......	‡ 1.01
" 29, "	30	128,000,000	1,575	58	96.3	......	0.59
" 30, "	30	128,000,000	1,400	198	85.9	94.9 (96.3)	0.58

* The count on Oct. 6 was made by Professor H. C. Ernst.
† One-half "Free Flow."
‡ One-half grain when samples were taken.
‖ Temperature of Applied Water 71°.

TABLE No. 8.

FILTRATION EXPERIMENTS.—MORISON MECHANICAL FILTER.

Growths of Water Bacteria of EIGHTY FIVE HOURS OR MORE AND END GROWTHS, *in all of the Samples of Filtered Water that were taken* THIRTY MINUTES OR LESS *after water commenced to flow from the filter.*

Date.	Minutes after Flow.	Gallons of Water Filtered per Acre, per 24 Hours.	Bacteria per Cubic Centimeter.		Per cent. of the Applied Bacteria Removed.	Average Percentage of the Applied Bacteria Removed.	Grains of Sulphate of Alumina used per Gallon.
			In Applied Water.	In Filtered Water.			
Oct. *6, 1893	20	131,000,000	7,795	60	99.2	99.2	0.60
Oct. 17, "	21	110,000,000	6,175+	82	98.7	......	†0.57
" 25, "	16	100,000,000	33,500	250	99.3	......	0.62
" 26, "	17	101,000,000	17,300	248	98.6	......	0.60
" 27, "	18	116,000,000	10,700	431	96.0	.....	0.61
" 28, "	19	106,000,000	7,200	268	96.3	......	0.81
" 30, "	18	106,000,000	1,700	66	96.1	......	0.56
Nov. 2, "	16	116,000,000	7,900	134	98.3	......	0.81
" 3, "	22	116,000,000	11,450	113	99.0	......	0.84
" 4, "	18	102,000,000	4,100	344	91.6	......	‡1.20
" 4, "	23	112,000,000	4,100	220	94.6	......	‡1.20
" 6, "	21	112,000,000	2,300	26	98.9	......	0.60
" 9, "	18	100,000,000	2,600	135	94.8	......	0.85
" 9, "	23	108,000,000	2,600	214	91.8	......	0.85
" 11, "	15	104,000,000	3,650	252	93.1	......	0.82
" 15, "	18	95,000,000	3,800	[illegible]	93.4	......	0.88

			Commenced to use Bacillus Prodigiosus.				
Nov. 23, 1893	21	99,000,000	15,850	536	96.6	......	0.60
" 24, "	30	132,000,000	14,000	1,103	92.1	......	0.59
Dec. 1, "	30	128,000,000	2,850	420	85.3	......	0.97
" 2, "	30	112,000,000	6,000	270	95.5	......	0.74
" 4, "	30	119,000,000	4,475	278	93.8	......	0.60
" 5, "	30	110,000,000	5,800	198	96.6	......	0.61
" 7, "	30	98,000,000	4,000	91	97.7	......	0.60
" 12, "	30	121,000,000	3,200	112	96.5	......	0.58
" 13, "	30	132,000,000	3,300	144	95.6	......	0.60
" 14, "	30	134,000,000	4,950	244	95.1	......	0.59
" 15, "	30	132,000,000	1,900	104	94.5	......	0.66
" 16, "	30	132,000,000	5,900	224	96.2	......	0.74
" 19, "	30	136,000,000	5,350	404	92.4	......	0.58
" 21, "	30	142,000,000	3,200	160	95.0	......	0.61
" 23, "	30	132,000,000	2,600	127	95.1	......	0.72
" 26, "	30	137,000,000	4,367	208	95.2	......	0.82
" 28, "	30	130,000,000	2,700	248	90.8	......	0.85
" 29, "	30	123,000,000	2,275	170	92.5	......	0.81
Jan. 1, 1894	30	128,000,000	2,533	708	72.0	......	0.82
" 2, "	30	128,000,000	2,850	358	87.4	......	0.85
" 3, "	30	136,000,000	3,375	304	91.0	......	0.84
" 4, "	30	148,000,000	5,025	202	96.0	......	0.85
" 5, "	30	134,000,000	3,775	226	94.0	......	0.82
" ‖ 6, "	30	136,000,000	2,700	256	90.5	......	1.00
" ‖ 8, "	30	137,000,000	4,000	430	89.3	92.7	0.58
			Ceased to use Bacillus Prodigiosus.			(92.7)	

TABLE NO. 8.—CONCLUDED.

DATE.	Minutes after Flow.	Gallons of Water Filtered per Acre, per 24 Hours.	Bacteria per Cubic Centimeter.		Per cent. of the Applied Bacteria Removed.	Average Percentage of the Applied Bacteria Removed.	Grains of Sulphate of Alumina used per Gallon.
			In Applied Water.	In Filtered Water.			
Jan. 9, 1894	30	132,000,000	3,400	156	95.4	......	0.60
" 10, "	30	128,000,000	1,725	132	92.3	......	0.84
" 11, "	30	121,000,000	2,150	100	95.3	......	0.61
" 12, "	30	134,000,000	875	94	89.3	......	0.81
" 13, "	30	128,000,000	1,633	106	93.5	......	0.72
" 15, "	30	134,000,000	1,600	116	92.8	......	0.84
" 16, "	30	134,000,000	775	296	61.8	......	0.58
" 17, "	30	113,000,000	3,375	350	89.6	......	0.82
" 18, "	30	121,000,000	3,800	234	93.8	......	0.59
" 19, "	30	132,000,000	2,767	192	93.1	......	0.83
" 20, "	30	125,000,000	5,200	208	96.0	......	0.72
" 22, "	30	134,000,000	11,200	682	93.9	......	0.85
" 23, "	30	128,000,000	4,133	514	87.6	90.3 (92.5)	0.80
			Washed filter-bed with Caustic Soda.				
Jan. 24, 1894	30	128,000,000	5,025	166	96.7	......	0.60
" 25, "	30	128,000,000	3,600	144	96.0	......	0.82
" 26, "	30	128,000,000	10,700	286	97.3	......	0.58
" 27, "	30	132,000,000	7,567	77	99.0	......	‡ 1.01
" 29, "	30	128,000,000	3,100	76	97.5	......	0.59
" 30, "	30	128,000,000	3,000	312	89.6	96.0 (96.8)	0.58

* The count on Oct. 6 was made by Professor H. C. Ernst.
† One-half "Free Flow."
‡ One-half grain when samples were taken.
‖ Temperature of Applied Water 71°.

TABLE NO. 9.

SUMMARY of the Average Percentages, of Applied Water Bacteria, that were REMOVED *by the Experimental Morison Mechanical Filter. Determined from Samples of Filtered Water that were taken* ONE HOUR OR MORE *after water commenced to flow from the filter. (See tables 2, 3, 4 and 5).*

REMARKS.	Samples that were taken the Same Hour as the Applied Water.		Samples that were taken One Hour or More after water commenced to flow from the filter.	
	End Growths.	Growths of about 90 Hours.	End Growths.	Growths of 85 Hours or More and End Growths.
Professor Ernst's counts.	Total number 5.	Total number 5.	Total number 5.	Total number 5.
July and Oct. 1893	99.5 (99.6)	99.5 (99.6)	99.5 (99.6)	99.5 (99.6)
	Total number 10.	Total number 10.	Total number 12.	Total number 16.
Oct. 17 to Nov. 11 and 18, 1893	99.0 (99.4)	99.2 (99.5)	98.6 (98.8)	98.8 (99.1)
	Total number 9.	Total number 9.	Total number 35.	Total number 128.
Nov. 23, 1893, to Jan. 8 and 9, 1894. Bacillus Prodigiosus used.	95.9 (96.8)	96.1 (96.9)	96.1 (96.9)	95.9 (96.8)

TABLE NO. 9.—CONCLUDED.

REMARKS.	Samples that were taken the Same Hour as the Applied Water.		Samples that were taken One Hour or More after water commenced to flow from the filter.	
	End Growths.	Growths of about 90 Hours.	End Growths.	Growths of 85 Hours or More and End Growths.
	Total number 13.	Total number 13.	Total number 58.	Total number 58.
Jan. 9 to Jan. 23, 1894....	91.2 (94.3)	94.6 (96.3)	92.8 (95.2)	92.8 (95.2)
	Total number 6.	Total number 6.	Total number 18.	Total number 18.
Jan. 24 to Jan. 30, 1894.................... After washing with Caustic Soda.	98.3 (98.5)	98.2 (98.5)	98.8 (98.9)	98.8 (98.9)
	Total number 38.	Total number 38.	Total number 123.	Total number 220.
Average of All the Above from Oct. 17 to Jan. 30..............................	95.5 (97.5)	96.7 (98.1)	95.2 (97.0)	95.5 (97.0)
	Total number 16.	Total number 16.	Total number 30.	Total number 34.
Average from Oct. 17 to Nov. 11 and 18, 1893, and from Jan. 24 to Jan. 30, 1894.......	98.7 (99.1)	98.8 (99.3)	98.7 (98.9)	98.8 (99.0)

TABLE NO. 10.

SUMMARY of the Average Percentages, of Applied Water Bacteria, that were REMOVED by the Experimental Morison Mechanical Filter. Determined from Samples of Filtered Water that were taken THIRTY MINUTES OR LESS after water commenced to flow from the filter. (See tables 6, 7 and 8).

REMARKS.	End Growths.	Growths of about 90 Hours.	Growths of 85 Hours or More and End Growths.
Professor Ernst's counts.	Total number 1.	Total number 1.	Total number 1.
Oct. 1893	99.2	99.2	99.2
	Total number 7.	Total number 8.	Total number 18.
Oct. 17 to Nov. 11 and 20, 1893	96.1 (96.9)	96.4 (97.4)	94.2 (96.9)
	Total number 13.	Total number 13.	Total number 25.
Nov. 23, 1893, to Jan. 8, 1894. Bacillus Prodigiosus used.	90.9 (92.7)	90.7 (92.5)	92.7 (92.7)

TABLE NO. 10.—CONCLUDED.

REMARKS.	End Growths.	Growths of about 90 Hours.	Growths of 85 Hours or More and End Growths.
	Total number 13.	Total number 13.	Total number 13.
Jan. 9 to Jan. 23, 1894	90.3 (92.5)	93.0 (95.1)	90.3 (92.5)
	Total number 6.	Total number 6.	Total number 6.
Jan. 24 to Jan. 30, 1894 After washing with Caustic Soda.	96.0 (96.8)	94.9 (96.3)	96.0 (96.8)
	Total number 39.	Total number 39.	Total number 62.
Average of All the Above from Oct. 17 to Jan. 30	92.4 (94.4)	93.1 (94.9)	92.9 (95.1)
	Total number 13.	Total number 13.	Total number 24.
Average from Oct. 17 to Nov. 11 and 20, 1893, and from Jan. 24 to Jan. 30, 1894	96.1 (96.8)	95.7 (97.1)	94.6 (96.9)

TABLE NO. 11.

Showing the number of times that Percentages of MORE THAN TWO PER CENT., *of the Applied Water Bacteria,* APPEARED *in the Filtered Water of the Morison Mechanical Filter. Determined from Samples of Filtered Water that were taken* ONE HOUR OR MORE *after water commenced to flow from the filter. Also the Percentages that the number of times are of the Total Number of Results obtained. (See tables 2, 3, 4 and 5).*

PER CENTS.	From Oct. 17 to Nov. 11 and 18, 1893, and from Jan. 24 to Jan. 30, 1894.							
	Samples taken the Same Hour as the Samples of Applied Water.				Samples taken One Hour or More after water commenced to flow from the filter.			
	End Growths.		Growths of about 90 Hours.		End Growths.		Growths of 85 Hours or More and End Growths.	
	Total number 16.		Total number 16.		Total number 30.		Total number 34.	
	Number of Times.	Per cent. of Total.	Number of Times.	Per cent. of Total.	Number of Times.	Per cent. of Total.	Number of Times.	Per cent. of Total.
3	2	12.5		....	2	6.7	2	5.9
4	1	6.3	....	....	2	6.7	2	5.9
5	....	....	1	6.3	....	....	....	....
Totals........	3	18.8	1	6.3	4	13.4	4	11.8

TABLE No. 12.

Showing the number of times that Percentages of the Applied Water Bacteria REMOVED *(by the Morison Mechanical Filter), were* ONE PER CENT. AND MORE LESS THAN THE AVERAGE PER CENT. REMOVED. *Determined from Samples of Filtered Water that were taken* ONE HOUR OR MORE *after water commenced to flow from the filter. Also the Percentages that the number of times are of the Total Number of Results obtained. (See tables 2, 3, 4 and 5).*

PER CENTS LESS THAN THE AVERAGE.	From Oct. 17 to Nov. 11 and 18, 1893, and from Jan. 24 to Jan. 30, 1894.							
	Samples taken the Same Hour as the Samples of Applied Water.				Samples taken One Hour or More after water commenced to flow from the filter.			
	End Growths. Average 98.7.		Growths of about 90 Hours. Average 98.8.		End Growths. Average 98.7.		Growths of 85 Hours or More and End Growths. Average 98.8.	
	Total number 16.		Total number 16.		Total number 30.		Total number 34.	
	Number of Times.	Per cent. of Total.	Number of Times.	Per cent. of Total.	Number of Times.	Per cent. of Total.	Number of Times.	Per cent. of Total.
1	2	12.5	1	6.3	2	6.7	2	5.9
2	1	6.3	....	....	1	3.3	1	2.9
3	1	6.3	....	....	2	6.7	2	5.9
4	....	....	1	6.3	....	....	....	....
Totals........	4	25.1	2	12.6	5	16.7	5	14.7

TABLE NO. 13.

Showing the number of times that Percentages of the Applied Water Bacteria REMOVED *(by the Morison Mechanical Filter), were* MORE THAN TWO PER CENT. LESS THAN THE AVERAGE PER CENT. REMOVED. *Determined from Samples of Filtered Water that were taken* ONE HOUR OR MORE *after water commenced to flow from the filter. Also the Percentages that the number of times are of the Total Number of Results obtained. (See tables 2, 3, 4 and 5).*

PER CENTS LESS THAN THE AVERAGE.	From Oct. 17 to Nov. 11 and 18, 1893, and from Jan. 24 to Jan. 30, 1894.							
	Samples taken the Same Hour as the Samples of Applied Water.				Samples taken One Hour or More after water commenced to flow from the filter.			
	End Growths. Average 98.7.		Growths of about 90 Hours. Average 98.8.		End Growths. Average 98.7.		Growths of 85 Hours or More and End Growths. Average 98.8.	
	Total number 16.		Total number 16.		Total number 30.		Total number 34.	
	Number of Times.	Per cent. of Total.	Number of Times.	Per cent. of Total.	Number of Times.	Per cent. of Total.	Number of Times.	Per cent. of Total.
3	1	6.3	....	....	2	6.7	2	5.9
4	....	....	1	6.3		....		....
Totals........	1	6.3	1	6.3	2	6.7	2	5.9

TABLE NO. 14.

Showing the number of times that Percentages of the Applied Water Bacteria REMOVED (by the Morison Mechanical Filter), were ONE PER CENT. AND MORE LESS THAN THE AVERAGE PER CENT. REMOVED. Determined from Samples of Filtered Water that were taken ONE HOUR OR MORE after water commenced to flow from the filter. Also the Percentages that the number of times are of the Total Number of Results obtained. (See tables 2, 3, 4 and 5).

PER CENTS LESS THAN THE AVERAGE.	From Nov. 23, 1893, to Jan. 8 and 9, 1894. (Bacillus Prodigiosus used).							
	Samples taken the Same Hour as the Samples of Applied Water.				Samples taken One Hour or More after water commenced to flow from the filter.			
	End Growths. Average 95.9. Total number 9.		Growths of about 90 Hours. Average 96.1. Total number 9.		End Growths. Average 96.1. Total number 35.		Growths of 85 Hours or More and End Growths. Average 95.9. Total number 128.	
	Number of Times.	Per cent. of Total.	Number of Times.	Per cent. of Total.	Number of Times.	Per cent. of Total.	Number of Times.	Per cent. of Total.
1	....	....	....	....	1	2.9	3	2.3
2	2	22.2	1	11.1	3	8.6	6	4.7
3	....	....	....	....	1	2.9	4	3.1
4	....	....	1	11.1	....	....	1	0.8
5	1	11.1	....	....	2	5.7	2	1.6
6	....	....	....	....	1	2.9	3	2.3
8	....	....	....	....	....	....	2	1.6
9	....	....	....	....	1	2.9	1	0.8
10	....	....	....	....	....	....	1	0.8
15	...	....	....	....	....	....	1	0.8
19	....	....	....	....	....	....	1	0.8
26	....	....	....	....	....	....	1	0.8
Totals........	3	33.3	2	22.2	9	25.9	26	20.4

Table No. 14.—Concluded.

Per Cents Less than the Average.	From Jan. 9 to Jan. 23, 1894.							
	Samples taken the Same Hour as the Samples of Applied Water.				Samples taken One Hour or More after water commenced to flow from the filter.			
	End Growths. Average 91.2.		Growths of about 90 Hours. Average 94.6.		End Growths. Average 92.8.		Growths of 85 Hours or More and End Growths. Average 92.8.	
	Total number 13.		Total number 13.		Total number 58.		Total number 58.	
	Number of Times.	Per cent. of Total.	Number of Times.	Per cent. of Total.	Number of Times.	Per cent. of Total.	Number of Times.	Per cent. of Total.
1	....	....	....	....	....	....	....	....
2	....	....	....	....	....	....	....	....
3	1	7.7	....	....	1	1.7	1	1.7
4	....	....	1	7.7	2	3.4	2	3.4
6	....	....	1	7.7	....	....	....	....
7	....	....	....	....	....	....	....	....
8	....	....	....	....	1	1.7	1	1.7
9	...	....	....	....	1	1.7	1	1.7
10	....	....	1	7.7	1	1.7	1	1.7
11	....	....	....	....	2	3.4	2	3.4
12	....	....	....	....	1	1.7	1	1.7
13	....	....	....	....	1	1.7	1	1.7
14	1	7.7	....	....	1	1.7	1	1.7
16	....	....	....	....	1	1.7	1	1.7
23	1	7.7	....	....	....	....	....	....
24	....	....	....	....	1	1.7	1	1.7
Totals........	3	23.1	3	23.1	13	22.1	13	22.1

TABLE No. 15.

Showing the number of times that Percentages of MORE THAN TWO PER CENT., of the Applied Water Bacteria, APPEARED in the Filtered Water of the Morison Mechanical Filter. Determined from Samples of Filtered Water that were taken THIRTY MINUTES OR LESS after water commenced to flow from the filter. Also the Percentages that the number of times are of the Total Number of Results obtained. (See tables 6, 7 and 8).

PER CENTS.	From Oct. 17 to Nov. 11 and 20, 1893, and from Jan. 24 to Jan. 30, 1894.					
	End Growths.		Growths of about 90 Hours.		Growths of 85 Hours or More and End Growths.	
	Total number 13.		Total number 13.		Total number 24.	
	Number of Times.	Per cent. of Total.	Number of Times.	Per cent. of Total.	Number of Times.	Per cent. of Total.
3	3	23.1	2	15.4	3	12.5
4	3	23.1	2	15.4	4	16.7
5	....	....	2	15.4	2	8.3
6	....	....	1	7.7	1	4.2
7	1	7.7	1	7.7	2	8.3
8	1	7.7	....	....	2	8.3
10	1	7.7	....	....	1	4.2

TABLE NO. 16.

Showing the number of times that Percentages of the Applied Water Bacteria REMOVED *(by the Morison Mechanical Filter), were* ONE PER CENT. AND MORE LESS THAN THE AVERAGE PER CENT. REMOVED. *Determined from Samples of Filtered Water that were taken* THIRTY MINUTES OR LESS *after water commenced to flow from the filter. Also the Percentages that the number of times are of the Total Number of Results obtained. (See tables 6, 7 and 8).*

PER CENTS LESS THAN THE AVERAGE.	From Oct. 17 to Nov. 11 and 20, 1893, and from Jan. 24 to Jan. 30, 1894.					
	End Growths. Average 96.1.		Growths of about 90 Hours. Average 95.7.		Growths of 85 Hours or More and End Growths. Average 94.6.	
	Total number 13.		Total number 13.		Total number 24.	
	Number of Times.	Per cent. of Total.	Number of Times.	Per cent. of Total.	Number of Times.	Per cent. of Total.
1	....	....	1	7.7	1	4.2
2	....	....	1	7.7	1	4.2
3	1	7.7	1	7.7	2	8.3
5	1	7.7	....	....	1	4.2
7	1	7.7	....	....	1	4.2
10	....	....	1	7.7	....	....
22	....	....	....	....	1	4.2
Totals	3	23.1	4	30.8	7	29.3

TABLE NO. 17.

Showing the number of times that Percentages of the Applied Water Bacteria REMOVED *(by the Morison Mechanical Filter), were* MORE THAN TWO PER CENT. LESS THAN THE AVERAGE PER CENT. REMOVED. *Determined from Samples of Filtered Water that were taken* THIRTY MINUTES OR LESS *after water commenced to flow from the filter. Also the Percentages that the number of times are of the Total Number of Results obtained. (See tables 6, 7 and 8).*

PER CENTS LESS THAN THE AVERAGE.	From Oct. 17 to Nov. 11 and 20, 1893, and from Jan. 24 to Jan. 30, 1894.					
	End Growths. Average 96.1.		Growths of about 90 Hours. Average 95.7.		Growths of 85 Hours or More and End Growths. Average 94.6.	
	Total number 13.		Total number 13.		Total number 24.	
	Number of Times.	Per cent. of Total.	Number of Times.	Per cent. of Total.	Number of Times.	Per cent. of Total.
3	1	7.7	1	7.7	2	8.3
5	1	7.7	....	....	1	4.2
7	1	7.7	....	....	1	4.2
10	....	....	1	7.7	....	....
22	....	....	....	....	1	4.2
Totals	3	23.1	2	15.4	5	20.9

TABLE No. 18.

Showing the number of times that Percentages of the Applied Water Bacteria REMOVED (by the Morison Mechanical Filter), were ONE PER CENT. AND MORE LESS THAN THE AVERAGE PER CENT. REMOVED. Determined from Samples of Filtered Water that were taken THIRTY MINUTES OR LESS after water commenced to flow from the filter. Also the Percentages that the number of times are of the Total Number of Results obtained. (See tables 6, 7 and 8).

PER CENTS LESS THAN THE AVERAGE.	From Nov. 23, 1893, to Jan. 8, 1894. (Bacillus Prodigiosus used).					
	End Growths. Average 90.9.		Growths of about 90 Hours. Average 90.7.		Growths of 85 Hours or More and End Growths. Average 92.7.	
	Total number 13.		Total number 13.		Total number 25.	
	Number of Times.	Per cent. of Total.	Number of Times.	Per cent. of Total.	Number of Times.	Per cent. of Total.
1	....	....	....	....	....	....
2	1	7.7	2	15.4	3	12.0
3	....	....	....	....	1	4.0
4	1	7.7	....	....	....	....
5	....	....	1	7.7	1	4.0
6	1	7.7	....	....	....	....
7	....	....	....	....	1	4.0
8	....	....	1	7.7	....	....
17	....	....	1	7.7	....	....
19	1	7.7	....	....	....	....
21	....	....	....	....	1	4.0
Totals	4	30.8	5	38.5	7	28.0

TABLE No. 18.—CONCLUDED.

Per Cents Less than the Average.	From Jan. 9 to Jan. 23, 1894.					
	End Growths. Average 90.3.		Growths of about 90 Hours. Average 93.0.		Growths of 85 Hours or More and End Growths. Average 90.3.	
	Total number 13.		Total number 13.		Total number 13.	
	Number of Times.	Per cent. of Total.	Number of Times.	Per cent. of Total.	Number of Times.	Per cent. of Total.
1	1	7.7	....	....	1	7.7
2	....	....	1	7.7	....	....
3	1	7.7	1	7.7	1	7.7
7	....	....	1	7.7	....	....
8	....	....	1	7.7	....	....
29	1	7.7	....	....	1	7.7
Totals	3	23.1	4	30.8	3	23.1

TABLE NO. 19.

FILTRATION EXPERIMENTS.—MORISON MECHANICAL FILTER.

Table showing the Percentage of Applied BACILLUS PRODIGIOSUS that was REMOVED from the water by filtration. Also the number of these Bacilli that were found in the Applied and Filtered Water and the length of time that they were Grown.

DATE.	Hour that Sample was taken.	Gallons of Water Filtered per Acre, per 24 Hours.	Bacillus Prodigiosus per Cubic Centimeter in Applied Water.			Bacillus Prodigiosus per Cubic Centimeter in Filtered Water.				
						End of Growth.		Last Growth obtained.		Per cent. of the Applied Bacilli Removed.
			Estimated by Flow.	From top of Filter.	Hours of Growth.	Number of Bacilli.	Hours of Growth.	Number of Bacilli.	Hours of Growth.	
Nov. 23, 1893.	9.00 A.M.	116,000,000	....	....	....	....	....	5	140	100.0
	10.00 A.M.	116,000,000	....	....	....	....	....	12	140	99.9
	11.00 A.M.	120,000,000	....	....	....	....	....	6	92	100.0
	12 M.	120,000,000	12,173	....	45	30	116	....	....	99.8
	1.00 P.M.	125,000,000	....	....	....	...	....	4	116	100.0
	2.00 P.M.	128,000,000	...	....	....	....	....	0	....	100.0
	3.00 P.M.	132,000,000	....	....	....	....	....	0	....	100.0
	3.40 P.M.	132,000,000	....	....	....	....	...	3	113	100.0
Nov. 24, 1893.	5.43 A.M.	128,000,000	....	....	....	....	....	13	117	99.9
Nov. 24, 1893.	* 8.42 A.M.	132,000,000	...	....	....	....	....	1	117	100.0
	9.00 A.M.	146,000,000	....	....	....	....	...	2	67	100.0
	10.00 A.M.	146,000,000	....	....	....	....	....	L	....	....
	11.00 A.M.	132,000,000	...	....	....	3	67	....	....	100.0
	12 M.	132,000,000	....	10,000,000	93	....	....	6	....	100.0

TABLE No. 19.—CONTINUED.

DATE.	Hour that Sample was taken.	Gallons of Water Filtered per Acre, per 24 Hours.	Bacillus Prodigiosus per Cubic Centimeter in Applied Water.			Bacillus Prodigiosus per Cubic Centimeter in Filtered Water.				
						End of Growth.		Last Growth obtained.		Per cent. of the Applied Bacilli Removed.
			Estimated by Flow.	From top of Filter.	Hours of Growth.	Number of Bacilli.	Hours of Growth.	Number of Bacilli.	Hours of Growth.	
Nov. 24, 1893.	1.00 P.M.	128,000,000	....	....	....	3	93	....	....	100.0
	2.00 P.M.	121,000,000	....	....	....	....	....	4	115	100.0
	3.00 P.M.	121,000,000	....	....	....	....	....	3	91	100.0
	3.40 P.M.	132,000,000	....	....	....	....	....	2	91	100.0
Nov. 25, 1893.	4.56 A.M.	128,000,000	....	....	...	....	....	791	....	100.0
Nov. 25, 1893.	9.00 A.M.	128,000,000	....	....	....	....	....	2	142	100.0
	10.00 A.M.	125,000,000	...	....	....	....	....	6	142	100.0
	11.00 A.M.	128,000,000	....	....	....	....	....	0	142	100.0
	12 M.	132,000,000	....	1,500,000	92	....	....	0	142	100.0
	2.00 P.M.	123,000,000	....	....	....	....	....	L	....	....
	3.00 P.M.	137,000,000	....	....	....	....	....	L	....	....
	3.40 P.M.	132,000,000	....	....	....	....	....	L	....	....
Nov. 28, 1893.	9.30 A.M.	128,000,000	....	....	....	....	....	0	93	100.0
	10.30 A.M.	132,000,000	....	....	....	....	....	0	93	100.0
	11.30 A.M.	132,000,000	....	....	....	....	....	0	93	100.0
	12.30 P.M.	128,000,000	....	5,010	93	....	....	0	93	100.0
	1.30 P.M.	125,000,000	....	....	....	....	....	0	67	100.0
	2.30 P.M.	128,000,000	....	....	....	....	....	0	67	100.0
	3.30 P.M.	125,000,000	....	....	....	....	....	L	....	....
Nov. 29, 1893.	4.45 A.M.	125,000,000	....	....	....	....	....	0	97	100.0

Dec. 1, 1893.	9.30 A.M.	136,000,000	....	....	....	8	166	....	....	99.9
	10.30 A.M.	140,000,000	....	....	....	6	142	....	....	100.0
	11.30 A.M.	136,000,000	....	....	....	....	....	29	190	99.8
	12.30 P.M.	132,000,000	11,900	....	190	....	....	2	142	100.0
	1.30 P.M.	128,000,000	....	....	....	....	....	6	190	100.0
	2.30 P.M.	134,000,000	....	....	....	....	....	2	190	100.0
	3.40 P.M.	132,000,000	....	....	....	2	190	....	....	100.0
Dec. 2, 1893.	9.30 A.M.	128,000,000	....	....	....	....	....	0	137	100.0
	10.30 A.M.	125,000,000	....	....	....	....	....	0	185	100.0
	11.30 A.M.	125,000,000	....	....	....	....	....	0	185	100.0
	12.30 P.M.	128,000,000	....	52	164	....	....	0	185	100.0
	1.30 P.M.	125,000,000	....	....	....	....	....	0	185	100.0
	2.30 P.M.	125,000,000	....	....	....	....	....	0	185	100.0
	3.40 P.M.	125,000,000	....	....	....	....	....	0	185	100.0
Dec. 5, 1893.	9.30 A.M.	123,000,000	....	....	....	1	119	....	....	100.0
	10.30 A.M.	125,000,000	....	....	....	....	....	2	95	100.0
	11.30 A.M.	128,000,000	....	....	....	....	....	0	142	100.0
	12 M.	128,000,000	....	6,365	....	....	....	....	....	....
	12.30 P.M.	128,000,000	....	....	....	1	95	....	....	100.0
	1.30 P.M.	128,000,000	....	....	....	1	90	....	....	100.0
	2.30 P.M.	125,000,000	....	....	....	....	....	1	90	100.0
	3.40 P.M.	125,000,000	....	....	....	....	....	1	114	100.0
Dec. 6, 1893.	9.30 A.M.	134,000,000	....	...	....	3	95	....	....	100.0
	10.30 A.M.	128,000,000	....	....	....	1	95	....	....	100.0
	11.30 A.M.	125,000,000	....	....	....	3	119	....	....	100.0
	12.30 P.M.	125,000,000	....	12,000	119	....	....	3	133	100.0
	1.30 P.M.	125,000,000	....	....	....	....	....	0	139	100.0

Table No. 19.—Continued.

Date.	Hour that Sample was taken.	Gallons of Water Filtered per Acre, per 24 Hours.	Bacillus Prodigiosus per Cubic Centimeter in Applied Water.			Bacillus Prodigiosus per Cubic Centimeter in Filtered Water.				
						End of Growth.		Last Growth obtained.		Per cent. of the Applied Bacilli Removed.
			Estimated by Flow.	From top of Filter.	Hours of Growth.	Number of Bacilli.	Hours of Growth.	Number of Bacilli.	Hours of Growth.	
Dec. 6, 1893.	2.30 P.M.	128,000,000	....		....	....	....	2	115	100.0
	3.40 P.M.	125,000,000	....		....	....	....	3	115	100.0
Dec. 12, 1893.	9.30 A.M.	128,000,000	....	....	....	....	....	0	90	100.0
	10.30 A.M.	125,000,000	....	....	....	....	....	0	116	100.0
	11.30 A.M.	128,000,000	....	....	....	....	....	0	90	100.0
	12.30 P.M.	128,000,000	....	1,000	116	....	....	2	116	100.0
	1.30 P.M.	125,000,000	....	....	....	....	....	0	90	100.0
	2.30 P.M.	131,000,000	....	....	....	....	....	0	90	100.0
	3.40 P.M.	128,000,000	....	....	....	....	....	0	116	100.0
Dec. 13, 1893.	9.30 A.M.	134,000,000	....	....	....	....	....	0	92	100.0
	10.30 A.M.	134,000,000	....	....	....	....	....	0	92	100.0
	11.30 A.M.	137,000,000	....	....	....	....	....	0	92	100.0
	12.30 P.M.	134,000,000	....	1,200	92	....	....	0	113	100.0
	1.30 P.M.	134,000,000	....	....	....	....	....	0	111	100.0
	2.30 P.M.	130,000,000	....	....	....	....	....	0	111	100.0
	3.40 P.M.	132,000,000	....	....	....	....	....	0	92	100.0
Dec. 14, 1893.	9.30 A.M.	136,000,000	....	....	....	....	....	0	113	100.0
	10.30 A.M.	137,000,000	....	....	....	....	....	0	113	100.0
	11.30 A.M.	136,000,000	....	....	....	....	....	0	113	100.0

Dec. 14, 1893.	12.30 P.M.	134,000,000	....	1,200	113	....	....	0	113	100.0
	1.30 P.M.	134,000,000	...	....	....	....	....	0	113	100.0
	2.30 P.M.	134,000,000	....	....	....	....	....	0	113	100.0
	3.40 P.M.	132,000,000	....	....	....	....	....	0	137	100.0
Dec. 15, 1893.	9.30 A.M.	137,000,000	....	....	....	....	....	0	137	100.0
	10.30 A.M.	132,000,000	....	....	....	1	89	....	....	100.0
	11.30 A.M.	132,000,000	....	....	....	2	89	....	....	100.0
	12.30 P.M.	134,000,000	....	5,000	113	1	113	....	....	100.0
	1.30 P.M.	130,000,000	....	....	....	1	113	....	....	100.0
	2.30 P.M.	128,000,000	....	....	....	....	....	0	137	100.0
	3.40 P.M.	132,000,000	....	....	....	....	....	0	137	100.0
Dec. 16, 1893.	4.35 A.M.	130,000,000	....	....	....	....	....	0	137	100.0
Dec. 16, 1893.	9.23 A.M.	132,000,000	....	....	....	....	....	0	137	100.0
	10.00 A.M.	132,000,000	....	....	....	....	....	12	89	99.7
	11.00 A.M.	132,000,000	....	....	....	....	....	0	89	100.0
	12 M.	134,000,000	....	4,000	137	....	....	22	137	99.5
	1.00 P.M.	132,000,000	....	....	....	....	...	6	113	99.9
	2.00 P.M.	131,000,000	....	....	....	3	111	....	....	99.9
	3.00 P.M.	134,000,000	....	....	....	1	111	....	....	100.0
	3.40 P.M.	131,000,000	....	...	....	1	87	....	....	100.0
Dec. 18, 1893.	9.30 A.M.	131,000,000	....	....	....	....	....	0	65	100.0
	10.30 A.M.	132,000,000	....	....	....	....	....	0	65	100.0
	11.30 A.M.	132,000,000	....	....	....	....	....	0	65	100.0
	12.30 P.M.	131,000,000	....	8,500	65	....	....	0	65	100.0
	1.30 P.M.	128,000,000	....	....	....	....	....	0	63	100.0
	2.30 P.M.	131,000,000	....	....	....	....	....	0	63	100.0
	3.40 P.M.	128,000,000	....	....	....	....	....	0	63	100.0

TABLE NO. 19.—CONTINUED.

DATE.	Hour that Sample was taken.	Gallons of Water Filtered per Acre, per 24 Hours.	Bacillus Prodigiosus per Cubic Centimeter in Applied Water.			Bacillus Prodigiosus per Cubic Centimeter in Filtered Water.				
						End of Growth.		Last Growth obtained.		Per cent. of the Applied Bacilli Removed.
			Estimated by Flow.	From top of Filter.	Hours of Growth.	Number of Bacilli.	Hours of Growth.	Number of Bacilli.	Hours of Growth.	
Dec. 19, 1893.	4.40 A.M.	128,000,000	....		....	....	....	0	63	100.0
Dec. 19, 1893.	9.30 A.M.	147,000,000	....	....	....	....	....	0	65	100.0
	10.30 A.M.	147,000,000	....	....	....	....	....	0	89	100.0
	11.30 A.M.	147,000,000	....	....	...	1	89	....	....	100.0
	12.30 P.M.	134,000,000	....	10,500	65	....	....	0	41	100.0
	1.30 P.M.	134,000,000	...	....	....	....	....	0	65	100.0
	2.30 P.M.	128,000,000	....	....	....	....	....	0	65	100.0
	3.40 P.M.	132,000,000	....	....	....	....	....	0	65	100.0
Dec. 20, 1893.	5.35 A.M.	132,000,000	...	....	....	....	....	0	65	100.0
Dec. 20, 1893.	9.30 A.M.	132,000,000	....	....	....	....	....	40	66	100.0
	10.30 A.M.	136,000,000	....	....	....	....	....	31	66	100.0
	11.30 A.M.	136,000,000	....	....	....	....	....	29	66	100.0
	12.30 P.M.	132,000,000	....	195,000	42	....	....	31	66	100.0
	1.30 P.M.	128,000,000	....	....	....	....	....	16	90	100.0
	2.30 P.M.	130,000,000	....	....	....	....	....	10	90	100.0
	3.40 P.M.	128,000,000	....	1,512,000	40	....	....	15	66	100.0
Dec. 21, 1893.	5.30 A.M.	132,000,000	....	....	....	....	....	5	90	100.0
Dec. 21, 1893.	9.30 A.M.	156,000,000	....	....	....	....	....	2	90	100.0
	10.30 A.M.	138,000,000	....	....	....	....	....	4	90	100.0

Dec. 21, 1893.	11.30 A.M.	132,000,000	....	....	....	....	....	0	90	100.0
	12.30 P.M.	132,000,000	....	81,000	42	....	....	1	114	100.0
	1.30 P.M.	130,000,000	....	....	....	....	....	0	114	100.0
	2.30 P.M.	128,000,000	....	....	....	....	....	1	90	100.0
	3.40 P.M.	121,000,000	....	....	....	....	....	1	90	100.0
Dec. 22, 1893.	9.30 A.M.	134,000,000	....	....	....	....	....	52	162	100.0
	10.30 A.M.	137,000,000	....	....	....	....	....	80	162	100.0
	11.30 A.M.	134,000,000	....	....	....	....	....	6	66	100.0
	12.30 P.M.	134,000,000	....	473,000	42	....	....	64	162	100.0
	1.30 P.M.	132,000,000	....	....	....	....	....	44	160	100.0
	2.30 P.M.	130,000,000	....	....	....	....	....	0	88	100.0
	3.40 P.M.	130,000,000	....	....	....	....	....	4	112	100.0
Dec. 23, 1893.	5.35 A.M.	128,000,000	....	....	....	....	....	0	137	100.0
	*									
Dec. 23, 1893.	8.23 A.M.	119,000,000	....	....	....	....	....	19	161	100.0
	*									
	8.38 A.M.	132,000,000	....	....	....	....	....	1	161	100.0
	9.30 A.M.	146,000,000	....	....	....	7	137	....	....	100.0
	10.30 A.M.	142,000,000	....	....	....	....	....	5	113	100.0
	11.30 A.M.	136,000,000	....	....	....	....	....	24	161	100.0
	12.30 P.M.	134,000,000	....	35,000	42	....	....	44	161	100.0
	1.30 P.M.	132,000,000	....	....	....	....	....	16	161	100.0
	2.30 P.M.	132,000,000	....	....	....	....	....	8	113	100.0
	3.30 P.M.	132,000,000	....	....	....	....	....	10	208	100.0
Dec. 26, 1893.	9.30 A.M.	137,000,000	....	....	....	....	....	74	88	100.0
	12.30 P.M.	128,000,000	....	186,000	41	....	....	136	88	99.9
	1.30 P.M.	128,000,000	....	....	....	....	....	28	135	100.0
	3.40 P.M.	130,000,000	....	....	....	....	....	144	86	99.9

TABLE NO. 19.—CONTINUED.

DATE.	Hour that Sample was taken.	Gallons of Water Filtered per Acre, per 24 Hours.	Bacillus Prodigiosus per Cubic Centimeter in Applied Water.			Bacillus Prodigiosus per Cubic Centimeter in Filtered Water.				
						End of Growth.		Last Growth obtained.		Per cent. of the Applied Bacilli Removed.
			Estimated by Flow.	From top of Filter.	Hours of Growth.	Number of Bacilli.	Hours of Growth.	Number of Bacilli.	Hours of Growth.	
Dec. 28, 1893.	9.30 A.M.	146,000,000	....	....	....	....	....	30	161	99.9
	12.30 P.M.	132,000,000	....	46,000	89	....	....	14	161	100.0
	1.30 P.M.	136,000,000	....	....	....	....	....	100	137	99.8
	3.40 P.M.	132,000,000	....	....	....	....	....	32	161	99.9
Dec. 29, 1893.	3.35 A.M.	130,000,000	....	....	....	....	....	32	114	99.9
Dec. 29, 1893.	9.30 A.M.	132,000,000	....	....	....	....	....	32,434	42	66.0
	12.30 P.M.	130,000,000	....	95,500	42	....	....	Great excess of Prodigiosus. All liquefied.		
	1.30 P.M.	132,000,000	....	....	....	....	....			
	3.40 P.M.	130,000,000	....	....	....	....	....			
Jan. 1, 1894.	9.30 A.M.	132,000,000	....	....	....	....	....	176	112	99.8
	12.30 P.M.	130,000,000	....	86,000	64	....	....	458	112	99.5
	1.30 P.M.	132,000,000	....	....	....	....	....	218	112	99.7
	3.40 P.M.	130,000,000	....	....	....	....	....	442	112	99.5
Jan. 2, 1894.	3.35 A.M.	128,000,000	....	....	....	....	....	582	112	99.3
Jan. 2, 1894.	9.30 A.M.	147,000,000	....	....	....	....	....	26	113	100.0
	12.30 P.M.	132,000,000	....	68,000	41	....	....	2	137	100.0
	1.30 P.M.	132,000,000	....	....	....	....	....	174	89	100.0
	3.40 P.M.	130,000,000	....	....	....	....	....	2	137	100.0

Jan. 3, 1894.	9.30 A.M.	142,000,000	....	....	....	....	....	14	90	100.0
	12.30 P.M.	137,000,000	....	50,000	66	....	....	0	162	100.0
	1.30 P.M.	137,000,000	....	....	....	....	....	0	162	100.0
	3.40 P.M.	136,000,000	....	....	....	....	....	0	162	100.0
Jan. 4, 1894.	3.30 A.M.	128,000,000	....	....	....	....	....	34	161	99.9
	*									
Jan. 4, 1894.	8.27 A.M.	148,000,000	....	....	....	....	....	66	161	99.9
	9.30 A.M.	148,000,000	....	....	....	....	...	6	137	100.0
	12.30 P.M.	134,000,000	....	57,000	65	....	....	0	161	100.0
	1.30 P.M.	132,000,000	....	....	....	2	161	....	....	100.0
	3.40 P.M.	130,000,000	....	....	....	....	....	68	137	99.9
Jan. 5, 1894.	3.50 A.M.	130,000,000	....	....	....	2	161	....	....	100.0
Jan. 5, 1894.	9.30 A.M.	136,000,000	....	....	....	....	....	92	138	99.9
	12.30 P.M.	130,000,000	....	105,000	66	....	....	28	138	100.0
	1.30 P.M.	132,000,000	....	....	....	....	...	80	138	99.9
	3.40 P.M.	130,000,000	....	....	....	....	....	290	114	99.7
Jan. 6, 1894.	3.40 A.M.	125,000,000	....	....	....	....	....	30	113	100.0
Jan. †6, 1894.	12.30 P.M.	146,000,000	....	11,200	65	....	....	6	113	100.0
†	1.30 P.M.	138,000,000	....	....	....	....	....	4	87	100.0
†	3.40 P.M.	132,000,000	....	....	....	8	111	....	....	100.0
Jan. †8, 1894.	9.30 A.M.	142,000,000	....	....	....	....	....	0	138	100.0
†	12.30 P.M.	130,000,000	....	38,000	66	2	138	....	....	100.0
†	1.30 P.M.	128,000,000	....	....	....	....	....	0	138	100.0
†	3.40 P.M.	128,000,000	....	....	....	....	....	0	138	100.0

Average.. 99.8

(99.97)

Average not including 66 per cent. of Dec. 29.............................. 99.97

* Before Prodigiosus solution was turned on. † Temperature of Applied Water 71°. "L" in table signifies that sample liquefied.

TABLE No. 19.—CONCLUDED.

Rates of Alumina added per gallon of Applied Water, not including "Free Flow," during the runs when Bacillus Prodigiosus was added to the Applied Water. (The quantity of "Free Flow" was always the same).

DATE.	Grains of Sulphate of Alumina used per Gallon, not including "Free Flow."	Remarks.	DATE.	Grains of Sulphate of Alumina used per Gallon, not including "Free Flow."	Remarks.
Nov. 23, 1893.	0.53	Throughout run.	Dec. 19, 1893.	0.51	Throughout run.
" 24, "	0.52	" "	" 20, "	0.53	" "
" 25, "	0.54	" "	" 21, "	0.54	" "
" 28, "	0.54	" "	" 22, "	0.51	" "
Dec. 1, "	0.50	From 9.30 A.M. to 11.30 A.M.	" 23, "	0.52	" "
" 1, "	1.00	" 12.30 P.M. to end of run.	" 26, "	0.74	" "
" 2, "	0.50	" 9.30 A.M. to 11.30 A.M.	" 28, "	0.77	" "
" 2, "	0.75	" 12.30 P.M. to end of run.	" 29, "	0.73	" "
" 5, "	0.53	Throughout run.	Jan. 1, 1894.	0.75	" "
" 6, "	0.52	" "	" 2, "	0.77	" "
" 12, "	0.51	" "	" 3, "	0.76	" "
" 13, "	0.53	" "	" 4, "	0.77	" "
" 14, "	0.52	" "	" 5, "	0.74	" "
" 15, "	0.59	" "	" 6,* "	0.76	" "
" 16, "	0.51	" "	" 8,* "	0.51	" "
" 18, "	0.52	" "			

"Free Flow" was not included in the above as the rates of Alumina added were sometimes changed during a run in order to ascertain if the efficiency of the filter would be improved or decreased.

* Temperature of Applied Water 71°.

TABLE NO. 20.

Showing the Chemical Analyses, that were made at different times during the Filtration Experiments, of Samples of River and Applied Water and Filtered Water from the Morison Mechanical Filter, by Professor John H. Appleton of Brown University.

The larger figures signify parts (by weight) in one million parts of water, (by weight). The smaller figures signify grains per American gallon of water, (weighing 58,372.2 grains).

REMARKS.	Total Residue.	Organic and Volatile Matter.	Mineral Matter.	Albuminoid Ammonia.	Ready-formed Ammonia.	Oxide of Iron. (Fe_2O_3).	Oxide of Aluminum. (Al_2O_3).	Alumina, etc. (Fe_2O_3).
June 7, 1893.								
River Water.........	37.60 2.195	13.40 .782	24.20 1.413			.74 .043	.57 .033	
Filtered Water......	38.50 2.247	13.40 .782	25.10 1.465	....		.20 .012	.30 .017	....
July 25, 1893.								
Applied Water......				.30 .0175	.03 .0018			
Filtered Water......				.08 .0047	.01 .0006			
Aug. 15, 1893.								
River Water.........		....	....	.26	.02			
*Filtered Water.....	92. 5.370	32. 1.868	60. 3.502	.08 .0047	0 0			

TABLE No. 20.—CONTINUED.

The larger figures signify parts (by weight) in one million parts of water, (by weight). The smaller figures signify grains per American gallon of water, (weighing 58,372.2 grains).

REMARKS.	Total Residue.	Organic and Volatile Matter.	Mineral Matter.	Albuminoid Ammonia.	Ready-formed Ammonia.	Oxide of Iron. (Fe_2O_3).	Oxide of Aluminum. (Al_2O_3).	Alumina, etc. (Fe_2O_3).
Oct. 2, 1893.								
River Water.........	53. 3.094	17. .992	36. 2.101	.24 .0140	.06 .0035			
†Filtered Water......	60. 3.502	16. .934	44. 2.569	.08 .0046	0 0			....
Oct. 11, 1893.								
Applied Water......	60.	17.	43.					3.
Filtered Water —								
‡ 11.10 A.M.......	90.	17.	73.					10.
11.15 A.M.......	79.	18.	61.					6.
11.20 A.M.......	77.	17.	60.					3.5
§ 11.25 A.M.......	67.	14.	53.					2.

* At end of run.
† 16 Minutes after water commenced to flow from the filter.
‡ 13 Minutes after water commenced to flow from the filter.
§ 28 Minutes after water commenced to flow from the filter.

TABLE NO. 20.—CONCLUDED.

Sample of Pawtuxet River Water taken at Pettaconset Pumping Station, May 20, 1893, at 7 A.M.

	Parts by weight, per million of water, by weight.
Sand and insoluble in acid	2.81
Oxide of iron, Fe_2O_3	.65
Oxide of aluminum, Al_2O_3	.48
Lime, CaO	2.89
Magnesia, MgO	.68
Potash, K_2O	1.33
Soda, Na_2O	1.53
Sulphur trioxide, SO_3	1.83
Nitrogen pentoxide, N_2O_5	1.15
Carbon dioxide, CO_2, to form normal carbonate	2.36
Chlorine, Cl	2.52
	18.23
Subtract oxygen equivalent to chlorine found	.56
	17.67
Unaccounted for	.45
Amount found independently as total mineral matter	18.12

Carbon dioxide, expelled from water by boiling,—3.02.

The Above Results computed into the form of compounds.

	Parts by weight, per million of water, by weight.
Sand and insoluble in acid	2.81
Common salt, $NaCl$	2.89
Potassium sulphate, K_2SO_4	2.46
Calcium chloride, $CaCl_2$	1.20
Calcium sulphate, $CaSO_4$	1.19
Calcium nitrate, $Ca(NO_3)_2$	1.75
Calcium carbonate, $CaCO_3$	2.14
Magnesium carbonate, $MgCO_3$	1.43
Ferric oxide, Fe_2O_3	.65
Aluminic oxide, Al_2O_3	.48
Carbon dioxide, CO_2, combined, but in excess	.67
	17.67
Solid residue, unaccounted for	.45
Mineral residue, found by test	18.12

Carbon dioxide, expelled from water by boiling,—3.02.

Table No. 21.

Filtration Experiments.—Morison Mechanical Filter.

Color of the Water during each run of the filter. 0=Distilled Water. Range from 0 to 10.

Date.	Average color of Applied Water.		Color etc. of Filtered Water.							
			21 minutes after commenced to flow.		Average of daily observations after 21 minutes.		Average of night observations.		Maximum observation.	
	Day.	Night.	Color.	Per cent. Removed.	Color.	Per cent. Removed.	Color.	Per cent. Removed.	Day.	Night.
May 20, 1893........	....	....	0.	....	3.9	....	....	....	7.	....
" 22, "	....	....	0.	....	1.3	....	....	....	2.	....
" 23, "	....	....	0.	....	0.9	....	....	....	3.	....
June 12, "	....	....	1.	....	0.1	....	....	...	1.	....
" 13, "	....	....	1.	....	1.0	....	....	....	2.	....
" 14, "	....	....	0.	....	0.6	....	....	...	2.	....
" 15, "	....	....	0.	....	0.0	....	....	....	0.	....
" 17, "	....	....	0.	....	1.1	....	....	....	3.	....
" 19, "	....	....	0.	....	0.7	....	....	...	3.	....
" 20, "	....	....	0.	....	1.4	....	....	....	4.	....
" 28, "	....	....	2.	....	1.7	....	....	....	5.	....
Average from May 20 to June 28.	....	....	0.36	....	1.2	....	...	....	....	....
June 28, 1893........	....	....	...	...	2.0	....	5.2	....	2.	10.
" 29, "	+10.0	....	1.	90.0	1.5	85.0	....	....	3.	....

June 29 and 30, 1893..	....	....	....	....	....	....	2.7	....	...	5.
" 30, 1893.........	+10.0	...	2.	80.0	4.0	60.0	2.8	....	5.	4.
July 1, 1893.........	10.0	....	....	....	4.0	60.0	....	....	5.	....
" 3, "	10.0	....	1.	90.0	2.2	78.0	....	....	4.	....
" 5, "	10.0	....	1.	90.0	1.3	87.0	....	....	2.	....
" 6, "	+10.0	....	3.	70.0	3.6	64.0	3.3	....	6.	6.
" 8, "	10.0	....	1.	90.0	1.8	82.0	....	....	3.	....
" 11 and 12, 1893..	10.0	....	1.	90.0	0.6	94.0	5.9	....	2.	8.
" 19, 1893.........	+10.0	....	2.	80.0	1.2	88.0	3.0	....	2.	4.
" 20, "	+10.0	....	2.	80.0	1.4	86.0	4.7	....	3.	5.
" 24, "	....	....	2.	....	1.2	....	....	....	2.	....
" 25, "	+10.0	....	1.	90.0	1.4	86.0	....	....	3.	....
" 26, "	....	....	1.	....	1.3	....	....	....	2.	....
" 27, "	+10.0	....	1.	90.0	1.1	89.0	....	....	2.	....
Aug. 2, "	....	....	2.	....	1.8	....	....	....	3.	....
" 3, "	+10.0	....	1.	90.0	1.5	85.0	....	....	3.	....
" 5, "	+10.0	....	1.	90.0	1.0	90.0	....	....	1.	....
" 7, "	+10.0	....	1.	90.0	1.1	89.0	....	....	2.	....
" 8, "	+10.0	....	1.	90.0	1.9	81.0	....	....	3.	....
" 10, "	+10.0	....	1.	90.0	1.6	84.0	....	....	3.	....
" 11, "	+10.0	....	2.	80.0	1.9	81.0	....	....	4.	....
" 14, "	+10.0	....	1.	90.0	2.0	80.0	....	....	4.	....
" 16, "	+10.0	....	0.	100.0	1.5	85.0	....	....	4.	....
" 17, "	+10.0	....	1.	90.0	1.5	85.0	....	....	3.	....
" 23, "	6.0	....	....	....	1.0	83.3	....	....	3.	....
Sept. 6, "	4.3	....	1.	76.7	0.8	81.4	....	....	2.	....
" 7, "	5.0	....	0.	100.0	0.8	84.0	....	....	2.	....
" 8, "	4.5	....	1.	77.8	1.0	77.8	....	....	1.	....
" 9, "	4.8	....	0.	100.0	0.7	85.4	1.7	....	1.	2.
" 11, "	4.0	....	0.	100.0	0.9	77.5	1.6	....	2.	2.

TABLE NO. 21.—CONTINUED.

DATE.	Average color of Applied Water.		Color etc. of Filtered Water.							
			21 minutes after commenced to flow.		Average of daily observations after 21 minutes.		Average of night observations.		Maximum observation.	
	Day.	Night.	Color.	Per cent. Removed.	Color.	Per cent. Removed.	Color.	Per cent. Removed.	Day.	Night.
Sept. 27, 1893	4.0	....	0.	100.0	0.8	80.0	1.5	....	1.	2.
" 28, "	4.0	....	1.	75.0	1.0	75.0	0.9	....	1.	1.
" 29, "	5.0	....	0.	100.0	0.9	82.0	1.0	....	1.	1.
" 30, "	4.0	....	1.	75.0	1.2	70.0	0.6	....	2.	1.
Oct. 2, "	4.0	....	1.	75.0	0.9	77.5	1.2	....	1.	2.
" 3, "	4.5	....	1.	77.8	0.8	82.2	1.6	....	1.	2.
" 4, "	5.0	....	1.	80.0	1.0	80.0	1.0	....	1.	1.
" 5, "	4.2	....	1.	76.2	0.9	78.6	1.6	....	2.	3.
" 6, "	4.0	....	1.	75.0	0.9	77.5	1.0	....	2.	1.
" 11, "	5.0	....	1.	80.0	0.7	86.0	1.2	....	1.	2.
" 12, "	5.0	....	1.	80.0	0.9	82.0	1.2	....	1.	2.
" 17, "	5.0	....	0.	100.0	0.7	86.0	1.7	....	1.	2.
" 23, "	4.0	....	0.	100.0	0.6	85.0	....	....	1.	....
" 23, "	6.0	....	....	....	1.0	83.3	2.1	....	1.	3.
" 25, "	6.0	....	0.	100.0	0.9	85.0	1.3	....	1.	2.
" 26, "	6.1	....	1.	83.6	1.1	82.0	....	....	2.	....
Average from June 28 to Oct. 26.	7.3	....	0.98	87.1	1.4	81.4	2.1	....	....	....
Oct. 28, 1893	6.3	....	1.	84.1	1.8	71.4	1.0	....	3.	1

Oct. 30, 1893........	7.8	....	1.	87.2	1.1	85.9	2.0	....	2.	2.
" 31, "	7.0	....	1.	85.7	1.2	82.9	2.6	....	2.	3.
Nov. 1, "	6.0	....	1.	83.3	1.4	76.7	2.6	....	2.	3.
" 2, "	6.0	....	1.	83.3	1.1	81.7	2.5	....	2.	3.
" 3, "	6.3	....	1.	84.1	1.3	79.4	2.6	....	2.	3.
" 6, "	7.0	....	2.	71.4	1.5	78.6	....	....	2.	....
" 6, "	7.0	7.0	2.	71.4	2.9	58.6	3.5	50.0	2.	4.
" 7, "	7.0	7.0	....	....	1.4	80.0	3.9	44.3	3.	6.
" 8, "	7.0	7.0	2.	71.4	1.3	81.4	3.3	52.9	2.	5.
" 9, "	7.0	7.1	2.	71.4	1.1	84.3	1.3	81.7	2.	3.
" 10, "	7.0	6.3	1.	85.7	1.4	80.0	2.1	66.7	2.	3.
" 11, "	6.0	6.4	1.	83.3	1.4	76.7	2.5	60.9	2.	3.
" 13, "	6.0	6.0	1.	83.3	1.4	76.7	2.5	55.6	3.	3.
" 14, "	6.0	6.8	1.	83.3	1.5	76.6	3.3	52.2	2.	4.
" 15, "	6.0	6.0	1.	83.3	1.0	83.3	2.2	63.3	1.	3.
" 16, "	6.0	6.5	1.	83.3	1.3	78.2	1.8	72.3	2.	3.
" 17, "	5.9	6.0	1.	83.1	1.0	83.1	1.4	76.7	1.	2.
" 18, "	6.0	6.0	1.	83.3	1.3	78.3	2.8	53.3	2.	3.
" 20, "	5.0	5.0	1.	80.0	1.0	80.0	1.5	70.0	1.	2.
" 21, "	5.0	5.0	1.	80.0	1.0	80.0	1.1	78.0	1.	2.
" 22, "	6.0	6.0	1.	83.3	1.0	83.3	1.7	71.7	1.	2.
" 23, "	5.0	5.7	1.	80.0	1.1	78.0	2.7	52.6	2.	3.
" 24, "	5.0	5.0	1.	80.0	1.1	78.0	1.2	76.0	2.	4.
" 25, "	5.3	5.0	1.	81.1	1.3	75.9	2.5	50.0	2.	3.
" 27, "	5.0	5.0	1.	80.0	1.0	80.0	1.0	80.0	1.	1.
" 28, "	5.0	5.1	1.	80.0	1.1	78.0	2.0	60.8	2.	2.
" 29, "	5.0	5.0	1.	80.0	1.2	76.0	2.4	52.0	2.	3.
Dec. 1, "	5.0	5.0	1.	80.0	1.1	78.0	2.0	60.0	2.	2.
" 2, "	5.0	5.0	2.	60.0	1.1	78.0	1.5	70.0	2.	2.
" 4, "	5.0	5.0	1.	80.0	1.2	76.0	2.1	58.0	2.	3.

Table No. 21.—Concluded.

Date.	Average color of Applied Water.		Color etc. of Filtered Water.							
			21 minutes after commenced to flow.		Average of daily observations after 21 minutes.		Average of night observations.		Maximum observation.	
	Day.	Night.	Color.	Per cent. Removed.	Color.	Per cent. Removed.	Color.	Per cent. Removed.	Day.	Night.
Dec. 5, 1893	5.0	5.0	1.	80.0	1.2	76.0	1.7	66.0	2.	3.
" 6, "	5.0	6.0	1.	80.0	1.0	80.0	1.7	71.7	1.	2.
" 7, "	5.0	5.0	1.	80.0	1.3	74.0	1.5	70.0	2.	2.
" 8, "	5.0	5.0	1.	80.0	1.1	78.0	2.3	54.0	2.	3.
" 12, "	5.0	5.0	1.	80.0	1.2	76.0	2.3	54.0	2.	3.
" 13, "	5.3	6.0	1.	81.1	1.3	75.5	2.5	58.3	2.	3.
" 14, "	6.0	6.0	1.	83.3	1.4	76.7	2.4	60.0	2.	3.
" 15, "	6.0	6.0	1.	83.3	1.4	76.7	1.9	68.3	2.	3.
" 18, "	6.0	6.0	1.	83.3	1.1	81.7	1.2	80.0	2.	2.
" 19, "	6.0	6.0	1.	83.3	1.3	78.3	2.0	66.7	2.	2.
" 20, "	6.0	6.0	1.	83.3	1.1	81.7	1.4	76.7	2.	2.
" 21, "	6.0	6.0	1.	83.3	1.3	78.3	1.6	73.3	2.	2.
" 22, "	6.0	6.0	1.	83.3	1.3	78.3	1.7	71.7	2.	2.
" 26, "	5.0	5.0	1.	80.0	1.0	80.0	1.0	80.0	1.	1.
" 27, "	5.0	5.0	1.	80.0	1.0	80.0	1.0	80.0	1.	1.
" 28, "	5.0	5.0	1.	80.0	1.0	80.0	1.2	76.0	1.	2.
" 29, "	5.0	5.0	1.	80.0	1.0	80.0	1.0	80.0	1.	1.
Jan. 1, 1894	5.0	5.0	1.	80.0	1.0	80.0	1.0	80.0	1.	1.
" 2, "	5.0	5.0	1.	80.0	1.0	80.0	1.0	80.0	1.	1.
" 3, "	5.0	5.0	1.	80.0	1.0	80.0	1.0	80.0	1.	1.
" 4, "	5.0	5.0	1.	80.0	1.0	80.0	1.0	80.0	1.	1.

Jan. 5, 1894.	5.0	5.0	1.	80.0	1.0	80.0	1.0	80.0	1.	1.
" 8, "	5.0	5.0	1.	80.0	1.0	80.0	1.5	70.0	1.	2.
" 9, "	5.0	5.0	1.	80.0	1.1	78.0	2.0	60.0	2.	2.
" 10, "	5.0	5.0	1.	80.0	1.0	80.0	1.1	78.0	1.	2.
" 11, "	5.0	5.0	1.	80.0	1.3	74.0	2.0	60.0	2.	2.
" 12, "	5.0	5.0	1.	80.0	1.0	80.0	1.0	80.0	1.	1.
" 15, "	5.0	5.0	1.	80.0	1.0	80.0	1.1	78.0	1.	2.
" 16, "	5.0	4.8	1.	80.0	1.2	76.0	2.0	58.3	2.	2.
" 17, "	5.0	5.0	1.	80.0	1.0	80.0	1.1	78.0	1.	2.
" 18, "	5.0	5.0	1.	80.0	1.3	74.0	2.3	54.0	2.	3.
" 19, "	5.0	5.0	1.	80.0	1.0	80.0	1.0	80.0	1.	1.
" 20, "	5.0	5.0	1.	80.0	1.1	78.0	....	....	2.	....
" 22, "	4.5	4.0	1.	77.8	1.0	77.8	1.0	75.0	1.	1.
" 23, "	4.5	....	1.	77.8	1.0	77.8	....	....	1.	1.
" 24, "	4.0	4.0	1.	75.0	1.2	70.0	2.0	50.0	2.	2.
" 25, "	4.0	4.0	1.	75.0	1.0	75.0	1.0	75.0	1.	1.
" 26, "	4.0	4.0	1.	75.0	1.2	70.0	2.3	42.5	2.	3.
" 27, "	4.0	....	1.	75.0	1.0	75.0	....	....	1.	....
" 29, "	4.0	4.0	1.	75.0	1.2	70.0	2.5	37.5	2.	3.
" 30, "	4.0	4.0	1.	75.0	1.2	70.0	2.4	40.0	2.	3.
Average from Oct. 28, 1893, to Jan. 30, 1894	5.4	5.4	1.1	79.9	1.2	77.9	1.8	66.3	....	

TABLE NO. 22. — PART 1.

*Relating to the Filtration of Water at Lawrence, Massachusetts, compiled from data obtained from the Report of the State Board of Health of Massachusetts for the year 1892.**

Filters.	Kind of Material.	Depth of Material.	Effective size of Sand in Millimeters.	Loam layers. Thickness in Inches.	Loam layers. Depth below surface in Inches.	Average Rate of Filtration per Acre, in Gallons, per 24 Hours.	Kind of Filtration.	Per cent. of Water Bacteria Removed. By Daily Samples.	Per cent. of Water Bacteria Removed. By Total Numbers.	Remarks.
18A,	Sand.	5 ft. 3 in.	0.48	0	0	1,294,000	Intermittent.	(98.7) 98.7	(99.3) 99.2	Started September, 1889.
41,	Sand and Loam.	5 " 0 "	0.14	3½	9	1,596,000	Intermittent.	(99.2) 98.6	(99.7) 99.6	Constructed May 9, 1892.
33A,	Sand.	5 " 0 "	0.14	0	0	1,862,000	Continuous.	(98.9) 97.6	(99.5) 99.2	Started April 28, 1892.
34A,	Sand.	5 " 0 "	0.09	0	0	1,718,000	Continuous.	(99.1) 98.5	(99.7) 99.5	Constructed April 28, 1892.
35A,	Sand and Loam.	4 " 9½ "	0.20	1	12	1,423,000	Intermittent.	(99.2) 98.0	(99.5) 98.8	Started March 30, 1892.
36A,	Sand and Loam.	4 " 10 "	0.20	1	12	1,778,000	Continuous.	(98.6) 98.0	(99.4) 99.3	Started March 30, 1892.
37,	Sand.	5 " 1 "	0.20	0	0	1,663,000	Continuous.	(98.4) 98.0	(99.5) 99.3	Constructed April 18, 1892.
38,	Sand.	2 " 0 "	0.20	0	0	1,714,000	Continuous.	(97.9) 97.9	(99.3) 99.3	Started April 28, 1892.
39,	Sand.	1 " 0 "	0.20	0	0	1,733,000	Continuous.	(98.3) 98.2	(99.3) 99.2	Started April 28, 1892.
40,	Sand and Loam.	1 " 0 "	0.20	1	11	1,288,000	Continuous.	(97.9) 95.8	(99.4) 98.6	Constructed April 28, 1892.
42,	Sand.	1 " 1 "	0.20	0	0	2,252,000	Continuous.	(96.6) 95.8	(98.1) 97.8	Started October 29, 1892.

* See description of tables.

FILTERS AND TOTAL RESULTS.

26 days.

Per Cents.	18A. Total 92.		41. Total 126.		33A. Total 144.		34A. Total 143.		35A. Total 128.		36A. Total 145.		37. Total 144.		38. Total 145.		39. Total 145.		40. Total 115.		42. Total 26.	
	Number of Times.	Per cent. of Total.	Number of Times.	Per cent. of Total.	Number of Times.	Per cent. of Total.	Number of Times.	Per cent. of Total.	Number of Times.	Per cent. of Total.	Number of Times.	Per cent. of Total.	Number of Times.	Per cent. of Total.	Number of Times.	Per cent. of Total.	Number of Times.	Per cent. of Total.	Number of Times.	Per cent. of Total.	Number of Times.	Per cent. of Total.
1	11	12.0	10	7.9	12	8.3	4	2.8	10	7.8	13	9.0	9	6.3	18	12.4	18	12.4	8	7.0	6	23.1
2	14	15.2	7	5.6	16	11.1	10	7.0	12	9.4	4	2.8	7	4.9	11	7.6	12	8.3	10	8.7	6	23.1
3	2	2.2	4	3.2	5	3.5	3	2.1	3	2.3	2	1.4	4	2.8	10	6.9	11	7.6	3	2.6	3	11.5
4	2	2.2	1	0.8	..	..	1	0.7	..	..	1	0.7	..	..	7	4.8	2	1.4	5	4.3	2	7.7
5	1	1.1	2	1.6	1	0.7	..	..	..	..	2	1.4	2	1.4	4	2.8	2	1.4	3	2.6	..	..
6	..	..	..	..	2	1.4	..	..	1	0.8	2	1.4	..	..	..	..	2	1.4	1	0.9	1	3.8
7	..	..	2	1.6	1	0.7	1	0.7	1	0.8	1	0.7	1	0.7	2	1.4	1	0.7	1	0.9	..	..
8	1	1.1	..	..	..	..	1	0.7	1	0.8	2	1.4	2	1.4	..	..	..	..	..	..	..	..
9	..	..	..	..	..	..	..	..	..	..	..	..	..	..	1	0.7	2	1.4	..	..	..	..
10	..	..	1	0.8	2	1.4	1	0.7	..	..	2	1.4	1	0.7	..	..	3	2.1	2	1.7	1	3.8
15	2	2.2	..	..	..	..	1	0.7	..	..	2	1.4	1	0.7	3	2.1	4	2.8	2	1.7	2	7.7
20	..	..	..	..	1	0.7	1	0.7	..	..	1	0.7	1	0.7	..	..	..	..	..	..	..	..
25	..	..	..	..	..	..	2	1.4	..	..	..	..	2	1.4	..	..	..	..	2	1.7	1	3.8
30	..	..	..	..	..	..	1	0.7	..	..	..	..	..	..	..	..	..	..	..	..	..	..
35	..	..	..	..	..	..	..	..	..	..	1	0.7	..	..	..	..	..	..	..	..	..	..
40	..	..	..	..	..	..	..	..	..	..	..	..	..	..	..	..	..	..	..	..	..	..
45	..	..	..	..	..	..	..	..	..	..	..	..	2	1.4	..	..	..	..	..	..	..	..
50	..	..	..	..	..	..	..	..	..	..	..	..	..	..	..	..	..	..	..	..	..	..
60	..	..	..	..	2	1.4	..	..	..	..	..	..	..	..	..	..	..	..	2	1.7	..	..
70	..	..	1	0.8	..	..	..	..	1	0.8	1	0.7	..	..	..	..	..	..	1	0.9	..	..
80	..	..	..	..	1	0.7	..	..	1	0.8	..	..	..	..	1	0.7	..	..	1	0.9	..	..
Totals.	33	35.9	28	22.2	43	29.9	26	18.2	30	23.4	34	23.4	32	22.2	57	39.3	57	39.3	41	35.7	22	84.6

Table No. 22. — Part 3.

Showing the number of times, from June 1 to November 30, 1892, inclusive, that Percentages of the Applied Water Bacteria Removed *(by the Lawrence Experimental Filters), were* One Per Cent. and More Less than the Average Per Cent. Removed. *Also the Percentages that the number of times are of the Total Number of Results obtained.*

FILTERS AND TOTAL RESULTS.

Per Cents Less than the Average.	18A. Average 98.7 Total 92.		41. Average 98.6 Total 126.		33A. Average 97.6 Total 144.		34A. Average 98.5 Total 143.		35A. Average 98.0 Total 128.		36A. Average 98.0 Total 145.		37. Average 98.0 Total 144.		38. Average 97.9 Total 145.		39. Average 98.2 Total 145.		40. Average 95.8 Total 115.		42. (26 days.) Average 95.8 Total 26.	
	Number of Times.	Per cent. of Total.	Number of Times.	Per cent. of Total.	Number of Times.	Per cent. of Total.	Number of Times.	Per cent. of Total.	Number of Times.	Per cent. of Total.	Number of Times.	Per cent. of Total.	Number of Times.	Per cent. of Total.	Number of Times.	Per cent. of Total.	Number of Times.	Per cent. of Total.	Number of Times.	Per cent. of Total.	Number of Times.	Per cent. of Total.
1	2	2.2	4	3.2	..	..	3	2.1	2	1.6	1	0.7	2	1.4	7	4.8	4	2.8	3	2.6	..	..
2	1	1.1	1	0.8	..	..	..	..	..	..	1	0.7	..	..	7	4.8	3	2.1	1	0.9	1	3.8
3	2	2.2	1	0.8	3	2.1	1	0.7	..	..	2	1.4	2	1.4	4	2.8	1	0.7	1	0.9	..	..
4	1	1.1	2	1.6	..	..	..	..	1	0.8	2	1.4	..	..	..	..	2	1.4	..	..	..	..
5	..	..	1	0.8	1	0.7	..	..	1	0.8	1	0.7	1	0.7	2	1.4	1	0.7	..	..	..	..
6	..	..	1	0.8	..	..	1	0.7	1	0.8	2	1.4	2	1.4	..	..	1	0.7	1	0.9	..	..
7	1	1.1	..	..	1	0.7	1	0.7	..	..	..	..	..	..	1	0.7	3	2.1	..	..	..	..
8	..	..	..	..	1	0.7	..	..	..	..	..	..	1	0.7	..	..	1	0.7	1	0.9	1	3.8
9	..	..	1	0.8	..	..	..	..	..	..	1	0.7	..	..	..	..	1	0.7	..	..	..	..
10	..	..	..	..	..	..	2	1.4	..	..	2	1.4	1	0.7	1	0.7	2	1.4	2	1.7	1	3.8
15	2	2.2	..	..	1	0.7	..	..	..	..	2	1.4	..	..	2	1.4	2	1.4	..	..	1	3.8
20	..	..	..	..	..	..	2	1.4	..	..	..	..	2	1.4	..	..	..	..	2	1.7	1	3.8
25	..	..	..	..	..	..	1	0.7	..	..	..	..	1	0.7	..	..	..	..	..	..	..	..
30	..	..	..	..	..	..	1	0.7	..	..	1	0.7	..	..	..	..	..	..	..	..	..	..
35	..	..	..	..	..	..	..	..	..	..	..	..	..	..	..	..	..	..	..	..	..	..
40	..	..	..	..	..	..	..	..	..	..	..	..	1	0.7	..	..	..	..	..	..	..	..
45	..	..	..	..	..	..	..	..	..	..	..	..	1	0.7	..	..	..	..	..	..	..	..
50	..	..	..	..	..	..	..	..	..	..	..	..	..	..	..	..	..	..	..	..	..	..
60	..	..	..	..	2	1.4	..	..	..	..	1	0.7	..	..	..	..	..	..	3	2.6	..	..
70	..	..	1	0.8	1	0.7	..	..	1	0.8	..	..	..	..	..	..	..	..	..	..	..	..
80	..	..	..	..	..	..	..	..	1	0.8	..	..	..	..	1	0.7	..	..	1	0.9	..	..

TABLE NO. 22. — PART 4.

SUMMARY of Parts 2 and 3, showing the Percentages that the number of times, that MORE THAN TWO PER CENT., of the Applied Water Bacteria, APPEARED in the Filtered Water, are of the Total Number of Results obtained. Also the Percentages that the number of times, that Percentages of the Applied Water Bacteria Removed, that were MORE THAN TWO PER CENT. LESS THAN THE AVERAGE PER CENT. REMOVED, are of the Total Number of Results obtained.

Filters.	Depth of Filtering Medium.	Percentages that the number of times, that More than Two Per cent., of the Applied Water Bacteria, Appeared in the Filtered Water, are of the Total Number of Results obtained.	Percentages that the number of times, that Percentages of the Applied Water Bacteria Removed, that were More than Two Per cent. Less than the Average Per cent. Removed, are of the Total Number of Results obtained.	Total Number of Results obtained.
18A,	5 ft. 3 in.	8.7	6.5	92
41,	5 " 0 "	8.7	5.6	126
33A,	5 " 0 "	10.4	6.9	144
34A,	5 " 0 "	8.4	6.3	143
35A,	4 " 9½ "	6.3	3.9	128
36A,	4 " 10 "	11.7	9.7	145
37,	5 " 1 "	11.1	8.3	144
		Average 9.3	Average 6.7	
38,	2 " 0 "	19.3	7.6	145
39,	1 " 0 "	18.6	9.7	145
40,	1 " 0 "	20.0	9.6	115
42,	1 " 1 "	38.5	15.4	26

CONCLUSIONS.

As, has already been stated under the head of "Bacteriological Work," I consider the bacteriological results previous to October 17, with the exception of those of Professor H. C. Ernst, to be somewhat unreliable on account of the bacterial colonies not having been cultivated a sufficient length of time to reach their full growths and, also, as "Fifteen-per-cent Gelatin" was used the greater part of the time instead of "Ten-per-cent Gelatin," I did not make use of them, but in an incidental manner, in forming my conclusions relative to the efficiency of the Experimental Morison Mechanical Filter.

Subsequent investigations, however, which I have previously mentioned, that were made since October 17, for the purpose of determining, if possible, the difference that there would have been if the colonies of bacteria had been cultivated longer and "Ten-per-cent Gelatin" used the whole of the time, previous to October 17, lead me to think that the average efficiency of the filter for the removal of water bacteria, determined from the results of all the runs of the Morison Mechanical Filter previous to October 17, while Basic Sulphate of Alumina and "Free Flow" were being used, was very nearly 99.0 per cent.

This result is corroborated, to a certain extent, by check-counts that were made by Professor H. C. Ernst, of the Harvard University Medical School, on July 20 and 21, and October 3, 4, 5 and 6, which are given in detail in the Bacteriological tables.

It will be seen by the tables that the Average Result of the efficiency of the filter by the counts made by Professor H. C. Ernst in July and October was more than 99.0 per cent., and that from October 17 to November 11, the Average Result was about 99.0 per cent., and that after November 11, about the time when Bacillus Prodigiosus was applied, the efficiency of the filter commenced to decrease. I shall, therefore, assume that the average efficiency of the filter for the removal of water bacteria, from April 5, 1893, to November 11, 1893, while Basic Sulphate of Alumina and "Free Flow" were being used, was about 99.0 per cent., and, as will be seen by the tables, from November 23 to January 8, at the time Bacillus Prodigiosus was used (which will be referred to in detail hereafter), the Average Result was reduced to about 96.0 per cent.

The use of Bacillus Prodigiosus was discontinued on January 8, and from January 9 to 23, the filter was run in the ordinary way with the Applied Water in its normal condition, and as will be seen by the Bacteriological tables, the Average Result was still further reduced during this time to 92.8 per cent.

The filter-bed was steamed and boiled on December 7, and on December 11. On December 7, steam was injected through the wash-pipe and forced up through the bed for one hour with an applied pressure of about twenty pounds per square inch. The bed was then washed and a sufficient depth of water having been left in the filter, to thoroughly saturate the bed, the bed was boiled for more than one hour. On December 11, the filter having been run as usual since December 7, the filter-bed was again thoroughly saturated with water and boiled for one hour and fifty minutes. There was not any improvement in the bacterial results, however, after either of these procedures.

Investigations were also made to ascertain if the low temperature of the Applied Water had anything to do with the decreased efficiency of the filter. A number of experiments were made by adding Basic Sulphate of Alumina at the rate of one-half ($\frac{1}{2}$) grain per gallon to water that was maintained at the temperature of about 75° and about 36° in glass jars, and it was found that a perceivable flocculent precipitate formed much quicker in the water at a temperature of about 75° than it did at a temperature of about 36°. Two runs were then made with the filter (on January 6 and 8), the Applied Water being maintained at the average temperature of 71° by the injection of steam. The results obtained during these two runs did not indicate that there was any improvement in the efficiency of the filter.

I then came to the conclusion, after the use of Bacillus Prodigiosus had been discontinued and the filter run from January 9 to 23, with the Applied Water in its normal condition, that the quartz grains of the filter-bed had become covered with foreign matter, and that it was quite possible that there were accumulations of the same in other parts of the filter, upon which bacteria were feeding and growing, and that many of the bacteria that had appeared in the samples of Filtered Water since November 23, originated from this source.

As the abnormal condition of the filter-bed was first noticed a day or two after the first application of the Bacillus Prodigiosus, it is possible that the solutions containing the same had something to do with the above-mentioned condition of the filter. This is hardly probable, however, as on July 27 and August 17, and October 11 and 12, one (1) liter of bouillon containing Cruickshank's Bacillus was daily applied to the filter without having any apparent detrimental effect.

After consulting with Professor J. H. Appleton, and experimenting with several salts and acids upon small quantities of quartz taken from the filter-bed, it was found that a solution of one (1) part of Caustic Soda and twenty-four (24) parts of water would cleanse the samples of quartz very thoroughly, the quartz being bleached during the operation from a dark brown color, to a color which was very nearly as light as the original color of the quartz when it was first put into the filter. It was decided, owing to this discovery, to cleanse the filter with the above-mentioned solution. This was done by washing the filter and filter-bed, under a head of about twenty-five feet, in the same manner that was generally followed in washing the filter and filter-bed with water, with the exception that the filter and bed were washed three times with the same solution, the solution being drained off and pumped up into a tank at the end of each washing.

When the Caustic Soda solution was drained from the filter, at the end of each washing, it was done very slowly allowing the bed to soak in the solution from fifteen (15) to twenty (20) minutes.

After the filter was washed with the Caustic Soda solution, as will be seen by the Bacteriological tables, the average efficiency of the filler was increased from 92.8 per cent., which was the average from January 9 to 23, to 98.8 per cent.

The inefficient working of the filter from November 23 to January 23, may have been influenced to a certain extent by the Applied Water having been quite clear and almost entirely free from suspended matter, it averaging during this time, as may be seen by table No. 21, showing the Color during each run of the filter, about five (5), while from October 17 to November 11, the average Color was about seven (7). Therefore, as the best work of the filter before November 23, seems to have been done when there was more suspended matter in the Applied Water than there was between November 23 and January 23, it would seem to indicate that a small quantity of suspended matter in the Applied Water exerted a beneficial influence in forming the supplementary filtering medium at the upper part of the quartz-bed.

If this reasoning is correct, the filter would have done better work after it had been washed with the Caustic Soda solution, if there had been more suspended matter in the Applied Water, as, at this time it was remarkably clear, and its color was the least that it had been during the work, namely :—four (4).

The average efficiency of the filter for removing Water Bacteria during the entire time it was running, from October 17 to January 30 inclusive, as may be seen by table No. 9, was about 95.2 per cent. I do not consider 95.2 a fair average, however, as undoubtedly between November 23 and January 23, the quartz grains of the filter-bed were, more or less, covered with growths of bacteria, and it is quite possible that bacteria were propagating in the other parts of the filter.

I have, therefore, based my bacteriological conclusions, which are to follow, upon the work that was done from October 17 to November 11, when the filter, from a bacteriological standpoint, appeared to be in its normal condition, and from January 24 to January 30, after the bed had been washed with the solution of Caustic Soda and water, and again brought to its normal condition.

I have assumed in preparing the estimate, which will follow, of the cost of operating a large Mechanical Filter Plant, that it will be necessary to wash the filters of the plant once every six (6) months with a solution containing Caustic Soda. The time that really elapsed during the experiments with the Experimental Morison Mechanical Filter, after commencing to use Basic Sulphate of Alumina, before the filter appeared to contain growths of bacteria, was rather more than seven months, but as the filter was not running during the entire time, and in order to be on the safe side in making the estimate, I decided to assume six (6) months, as mentioned above.

The Average Percentages relative to the Morison Mechanical Filter, which I shall quote hereafter in connection with Bacteriological Work, will be the End Growths given in table No. 9, under the heading of "Samples that were taken One Hour or More after water commenced to flow from the filter," and the Average Percentages under the heading of "End Growths" given in table No. 10, relating to Samples taken Thirty Minutes or Less after water commenced to flow from the filter.

I will now sum up in detail the different points of investigation, relative to the Experimental Morison Mechanical Filter, which are mentioned on page 36.

First.—The chemicals best adapted for the purification of Pawtuxet River Water, viz.:—

Basic Sulphate of Alumina. The quality used contained from 15.8 to 17.5 per cent. of Alumina (Al_2O_3).

Second.—The best method of applying the chemicals and the quantity to add to the Applied Water for each gallon of water filtered, viz.:—

The method of doing this has already been described, and I think the experiments have demonstrated that Basic Sulphate of Alumina added to the Applied Water at the rate of one-half ($\frac{1}{2}$) grain per gallon, and "Free Flow" gave as good results as three-fourths ($\frac{3}{4}$) of a grain and "Free Flow," that we have repeatedly tried, and a larger quantity than three-fourths ($\frac{3}{4}$) and "Free Flow" that was tried in several instances, as will be seen in the tables. I shall, therefore, base my estimates upon six-tenths ($\frac{6}{10}$) of a grain of Alumina per gallon, including "Free Flow" (for an average run of 16 hours and 43 minutes), that being equivalent to one-half ($\frac{1}{2}$) grain per gallon while in effective service.

Third.—If any portion of the chemicals that were added to the Applied Water were present in the Filtered Water, viz.:—

The results, that I have mentioned, that were obtained by applying the Logwood and Acetic Acid test for Alum, in conjunction with filter-paper, have demonstrated, I think, that none of the Basic Sulphate of Alumina was present during the experiments in the Filtered Water, in its original state, after the water had been flowing from the filter twenty-one (21) minutes. The only indication of Alumina, found in the Filtered Water, was a minute quantity of finely suspended hydrate, resulting from the addition of the Alumina, that came through the filter-bed with the water that was being filtered.

It is also shown in table No. 20, by an analysis made June 7, 1893, that the Oxide of Aluminium or Alumina (Al_2O_3), found in the Filtered Water was forty-seven (47) per cent. less than it was in the River Water, and that on October 11, 1893, twenty-three (23) minutes after the water commenced to flow from the filter, the Alumina, etc. (Fe_2O_3), in the Filtered Water was practically the same as that in the Applied Water.

An analysis by Professor Thomas M. Drown, whose report is appended to this report, shows that 0.0292 of a grain of Alumina ($Al_2 O_3$), per gallon, was found in a sample of Pawtuxet River Water, that had been taken directly from the river and afterwards filtered through a double thickness of Swedish paper, and that 0.0584 of a grain of Alumina (Al_2O_3), per gallon, was found in a sample of the same water, after Sulphate of Alumina had been added to it, at the rate of one-half (½) grain per gallon, and the very slight flocculent precipitate produced filtered off through a double thickness of filter-paper, showing an increase of Alumina ($Al_2 O_3$), of 0.0292 of a grain.

Fourth.—The rate in gallons per Acre per 24 hours which could be efficiently filtered, viz.:—

The Bacteriological tables show that the water has been filtered successfully from a rate of 90,000,000 gallons to a rate of 193,000,000 gallons per Acre per 24 hours, the average rate of filtration being about 128,000,000. If a mechanical filter-plant should be constructed for the City of Providence, however, I should recommend that its average capacity be based upon a rate of 100,000,000 gallons per Acre per 24 hours, in order to have a sufficient reserve to insure the practical working of the plant while several filters are being washed at the same time, and to meet unforeseen contingencies which may arise in the future, etc., etc.

Fifth.—The Bacteriological and Chemical purification of the water, viz.:—

The following table gives the AVERAGE OF THE PERCENTAGES, from October 17 to November 11, 1893, and from January 24 to 30, 1894, given in tables from No. 2 to No. 8 inclusive, of Applied Water Bacteria, which were Removed by the filter, determined from Samples of Filtered Water, that were taken THIRTY MINUTES OR LESS AND ONE HOUR OR MORE, after water commenced to flow from the filter.

Including Samples taken *Thirty Minutes or Less* after water commenced to flow from the filter, and those taken at the *Same Hour as the Applied Water* (which was One Hour or More after water commenced to flow from the filter).		Including Samples taken *Thirty Minutes or Less*, and all *Samples taken One Hour or More*, after water commenced to flow from the filter.	
End Growths.	Growths of about 90 Hours.	End Growths.	Growths of 85 Hours or More and End Growths.
98.6	98.7	98.6	98.7

This table shows that the average efficiency of the filter for removing water bacteria was 98.6 per cent., (from October 17 to November 11, 1893, and from January 24 to January 30, 1894, when the filter-bed was apparently free from growths of bacteria).

A comparison of the Average Percentage of Removal of Applied Water Bacteria of the Experimental Morison Mechanical Filter, with the Average of Percentages of Removal computed from results obtained by Natural Filtration with the Experimental Filters at Lawrence, Mass., mentioned in the description of table No. 22, and referred to in the First part of table No. 22, as having beds about five (5) feet deep, is as follows:—

Morison Mechanical Filter.

Average Percentage of Removal........... 98.6

Natural Filters at Lawrence, Mass.

Average of the Average Percentages of Removal, given in the Ninth column of the first part of table No. 22, not inclosed in parentheses, 98.2

Morison Mechanical Filter............More 00.4 per cent.

Natural Filters at Lawrence, Mass.

Average of the Average Percentages of Removal, given in the Ninth column of the first part of table No. 22, inclosed in parentheses, 98.9

Morison Mechanical Filter............Less 00.3 per cent.

Natural Filters at Lawrence, Mass.

Average of the Average Percentages of Removal, given in the Tenth column of the first part of table No. 22, not inclosed in parentheses,	99.3
Morison Mechanical Filter ....Less	00.7 per cent.

Natural Filters at Lawrence, Mass.

Average of the Average Percentages of Removal, given in the Tenth column of the first part of table No. 22, inclosed in parentheses,	99.5
Morison Mechanical Filter ....Less	00.9 per cent.

I do not consider that the efficiency of a filter should be entirely based upon the average results obtained, although this is generally the standard upon which the efficiency is based, but that the worst results obtained should be duly considered. In order to present my ideas upon this subject more clearly I will assume a rather improbable case. For example, if one hundred individual results were used in working up an average, 90 of these results might each show an efficiency of 100 per cent., and 10 of them might each show an efficiency of only 80 per cent., or in other words 10 per cent. of the Total Results would be 18 per cent. below the Average Result, which in my opinion would be sufficient grounds to condemn a filter. Yet the average of the whole number would be 98.0 per cent., which is a very good result. The tables from No. 11 to No. 18 inclusive and table No. 22 were prepared in order to make comparisons of this kind.

It is shown in table No. 9, that the average efficiency of the Experimental Morison Mechanical Filter is 98.7 per cent. (computed from all of the results that were obtained from samples that were collected *One Hour or More* after water commenced to flow from the filter), and by inspecting table No. 12, it can be seen that 16.7 per cent. of the Total Results are from 1 to 3 per cent. Less than the Average Result, also by inspecting table No. 13, it can be seen that 6.7 per cent. of the Total Results are More than 2 per cent. Less than the Average Result, namely: 2 of 3 per cent. each. It is also shown in table No. 10, that the average efficiency of the filter,

Thirty Minutes or Less after water commenced to flow, is 96.1 per cent., and by inspecting table No. 16, it can be seen that 23.1 per cent. of the Total Results are from 1 to 7 per cent. Less than the Average Result, also by inspecting table No. 17, it can be seen that 23.1 per cent. of the Total Results are More than 2 per cent. Less than the Average Result, namely :—from 3 to 7 per cent.

The results given in tables Nos. 14 and 18 I have not considered in the above comparisons, as they cover periods during which the filter contained growths of bacteria (from November 23, 1893, to January 9, 1894, and from January 9 to January 23, 1894), relative to which I have previously referred several times.

For the purpose of comparing these per cents, just mentioned, with per cents obtained with Natural Filtration, I will call attention to table No. 22, which was computed from the Report of the Massachusetts State Board of Health for the year 1892.—It can be seen by examining the Third part of this table, by the percentages given, of filters having a depth of filter-bed of about five (5) feet (I shall not consider the filters having a depth of bed less than about five (5) feet, as I should not suppose they would be used in ordinary practice). that 8.7 per cent. of the Total Results are from 1 to 80 per cent. Less than the Average Result, and by examining the Fourth part of the table it can be seen that 6.7 per cent. of the Total Results are More than 2 per cent. Less than the Average Result, namely :—from 3 to 80 per cent.

The percentages show in the comparison of the One Hour or More results of the Morison Mechanical Filter, in the case of the One and more per cents, that the percentage of the number is in favor of the Natural Filters, namely: 8.7 per cent. less than the average, to 16.7 per cent. less than the average for the Morison Mechanical Filter; but, that the range of the per cents of removal is very largely in favor of the Morison Mechanical Filter, namely: 1 to 3 per cent. less than the average, to from 1 to 80 per cent. less than the average for the Natural Filters.—And in the case of the More than 2 per cents, that the percentage of the number is the same for both filters, namely: 6.7, but, that the range of the per cents of removal, is, as above, very largely in favor of the Morison Mechanical Filter, namely: 2 of 3 per cent. less than the average, to from 3 to 80 per cent. less than the average for the Natural Filters.

The percentages show, in the comparison of the Thirty Minutes or Less results of the Morison Mechanical Filter, as compared with the daily average results of the Natural Filters, that the percentages of the number are in both instances in favor of the Natural Filters, namely:—in the case of the One and more per cents, 8.7 per cent. less than the average, to 23.1 per cent. less than the average for the Morison Mechanical Filter, and in the case of the More than two per cents, 6.7 per cent. less than the average, to 23.1 per cent. less than the average for the Morison Mechanical Filter, but that the range of the per cents of removal is very largely in favor of the Morison Mechanical Filter, namely:—from 3 to 7 per cent. less than the average, to from 3 to 80 per cent. less than the average for the Natural Filters. It should be remembered, however, that the Thirty Minutes or Less results affect the general average of a run of the Morison Mechanical Filter but in a very slight degree, as in one-half hour later at the greatest, the results are at their best, namely:—those of One Hour or More. The value of the Thirty Minutes or Less results in computing an average percentage, considering the run of the filter as 16 hours and 43 minutes, which was the average run during the experiments, would be, as compared with the value of the average of the One Hour or More results, about one (1) for the Thirty Minutes or Less results to about thirty-four (34) for the One Hour or More results.

Before proceeding any further, I will again call attention to what has previously been mentioned in the description of table No. 22, namely:—that some of the bacteria, which were found in some of the samples that I have considered in computing the percentages used in making the comparisons, the Massachusetts Report states, appeared to have had their origin in the interior of the filters and in the outlet-pipes and underdrains. It should also be borne in mind, in regard to the comparisons, that there is a considerable difference in the number of results that were used in compiling the tables upon which the comparisons are based, namely:—for table No. 22, the number of results ranged from 92 to 145, and for tables Nos. 12, 13, 16 and 17, the number of results were 13 and 30. Attention is also called to the fact, that for reasons which I have previously mentioned in detail (on account of cold weather), the December bacterial results given in the Massachusetts Report have not been used in the discussions in that report, nor have they been tabulated or considered in the comparisons made in this report. Quite a number of the experiments, however, that were made with the Experimental Morison Mechanical Filter, as can be seen by the tables relative to the same, were made in December and January, the temperature of the building in which the filter was located being kept above the "freezing point." The approximate Mean Temperature of the Applied Water, of the Morison Mechanical Filter, was in December 35°, and in January 34°. The Lowest Temperature of the Applied Water during these two months was about 33°.

It might appear, without an explanation, as though I had not been consistent in the use of data in my bacteriological comparisons of what has been accomplished with the Experimental Morison Mechanical Filter and what has been accomplished by Natural Filtration with the Experimental Filters at Lawrence, as the only results of the Morison Mechanical Filter that were considered were those of from October 17 to November 11, 1893, and those of from January 24 to January 30, 1894, when this filter was in its normal condition, and which did not include the results that were obtained when the filter contained foreign matter upon which bacteria were propagating, while the results obtained by Natural Filtration at Lawrence, that were considered, included those in which the number of bacteria in the Filtered Water exceeded 500, the latter not being used at Lawrence, as has previously been stated, in working out the averages of the results of Natural Filtration which are given in the Massachusetts Report.

I took into consideration, however, our experience with the Experimental Morison Mechanical Filter, and, in order to express my views clearly in regard to the matter, will present the following: If a first class Mechanical Filter-plant, properly housed, should be built in accordance with the above mentioned experience, having a capacity of about 15,000,000 gallons per 24 hours, and capable of filtering at the rate of from 100,000,000 to 150,000,000 gallons per Acre per 24 hours, and it should be found at any time, that the Filtered Water of the plant contained a number of bacteria, largely in excess of the usual number, as was the case of that of the Experimental Morison Mechanical Filter during the period when bacteria were propagating in the interior of the filter, it would not, probably, be necessary to devote any time to theorizing as to the cause of the occurrence, as the filters of the plant could be thoroughly cleansed by washing them with a solution of Caustic Soda and water in a short time and at a very slight expense in proportion to the total cost of operating the plant. In actual practice, however, it might be advisable to cleanse the filters at stated periods, such as practical experience might demonstrate, and not to wait for an excessive number of bacteria to appear in the Filtered Water, and it is quite likely that the process of cleansing with Caustic Soda could be improved upon by heating the solution with steam, and that as good results could be attained with a cheaper chemical compound in connection with steam. Whereas, if, for instance, a system of Natural Filter-beds, having the same capacity as above mentioned and capable of filtering at the rate of from 2,000,000 to 3,000,000 gallons per Acre per 24 hours, should meet with an experience similiar to what took place in the interior of the Experimental Natural Filters at Lawrence during the warm summer months of 1892, which is mentioned in the Massachusetts Report and which has previously been mentioned in this report, namely:—that the Filtered Water at times contained a very large number of bacteria, which in some cases equalled and even exceeded the number applied and which appeared to have had their origin in the interior of the filters and in the outlet-pipes and underdrains, it would be a very difficult matter to free the filters of the abnormal growths of bacteria, even if it could be done by artificial means, but by the expenditure of a very considerable amount of time and money, on account of the large area of the filter-beds, which would be required.

Of course, if the excessive number of bacteria mentioned above were all of a harmless species their appearance in the Filtered Water, under ordinary circumstances, would not be a matter of particular moment, although the stability of the filter-beds might be questioned until it was proved by a careful investigation that the excessive number of bacteria originated in the interior of the filters; but if there should happen to be a few cases of typhoid fever prevalent within the boundaries of the water-shed of the supply from which the Unfiltered Water was derived, at a time that an excessive number of bacteria was found in the Filtered Water, and there was the slightest possibility of even a portion of the dejecta of a typhoid fever patient finding its way into the supply, it would be a matter of very grave concern, as the question would naturally arise as to whether there were "breaks" in the filter-beds or if something unusual had happened to them, thereby causing an apprehension that there was a possibility of typhoid bacilli finding their way into the Filtered Water, which apprehension, if it were known that the filter-beds were in a proper condition, would not be felt. (Quite a number of epidemics of cholera and typhoid fever are known to have occurred in Europe, owing to the freezing of the whole or a portion of the surface of open Natural Filter-beds during the process of "scraping"). It is very probable, that an expert bacteriologist who was perfectly familiar with the species of bacteria that were generally found in the Unfiltered and Filtered Water of the supply under consideration as well as to the appearance of the typhoid bacillus in the water, could, after a cultivation of about two days, which is the length of time which would probably be necessary for the majority of the ordinary water bacterial colonies to become visible, detect suspicious appearing bacilli, were any such present in the samples which were being examined; but it would very likely require, even if there were not any suspicious bacilli discovered at first, at least one week to absolutely prove that there were not any typhoid bacilli present in the samples, and a much longer time to obtain sufficient reliable data to demonstrate that the filter-beds were intact and that the excessive number of bacteria originated, if such was the case, in the interior of the filters, unless there had been similar experiences in the past that had been thoroughly investigated, which could be taken as criterions from which to theorize.

As will be seen, by inspecting the tables, the most unfavorable results were obtained with the Experimental Morison Mechanical Filter from fifteen (15) minutes to about thirty (30) minutes after water first commenced to flow from the filter. I have previously stated, that with nine exceptions, River Water was used in washing the filter. If Filtered Water had been used the entire time, instead of River Water, it is quite probable that the "Thirty-Minute or Less" Results would have been somewhat better, and it is quite probable that more satisfactory results would have been obtained before the end of Thirty Minutes. It is simply a question, however, of the decrease or the increase of this time (30 minutes), as our experiments have positively shown that One Hour after the water commenced to flow from the filter the results were as good as they were at any time during a run of the filter. I have assumed, in working up the estimate of cost relative to waste-water, that the Filtered Water will have to be run to waste for one-half hour after water commenced to flow from the filter.

Since one of the most important points in the filtration of water is the removal of disease producing germs, the filter was tested by the application of Bacillus Prodigiosus. It would have been very dangerous to have added typhoid fever or other disease germs to the Applied Water, as the filter was located at the City's source of water supply.

It is stated in the Report of the State Board of Health of Massachusetts for the year 1892, that it was found at the Experimental Station at Lawrence, Mass., that with Bacillus Prodigiosus, more results fully as reliable and under more nearly parallel conditions could be obtained than by working with typhoid fever germs. It was further determined by a series of experiments which were made at the above-mentioned station, that the life histories of Bacillus Prodigiosus and Bacillus Typhi Abdominalis in the Merrimac River at Lawrence are quite similar.

Table No. 19 shows that the average percentage, of the Applied Bacillus Prodigiosus, that was removed from the water by filtration, was 99.8 per cent. The lowest percentage of efficiency given in the table is that of December 29, when the percentage of efficiency, given in the table, of the only sample that could be counted on that day, on account of the colonies liquefying, was 66.0 per cent. The average percentage of efficiency of the filter, if this 66.0 per cent. was not considered, would be 99.97 per cent. I am not able to offer a satisfactory explanation relative to the very large amount of Bacillus Prodigiosus found in the samples taken on December 29, as the filter, so far as could be judged by observation, was working in the usual manner on this day. The Color was of the average standard, and the half-hourly readings of the depth of water upon the filter-bed, and the quantity of water flowing into the filter, which are shown on Diagram No. 1, tend to show that the filter was working properly. It is barely possible, though not probable, that the glass tubes, in which the samples of water were collected on this day, were not sufficiently sterilized, as a great amount of work was done daily in the laboratory by the bacteriologist. The bacterial colonies in about fifty-one dishes being counted daily, and all of the sample tubes and dishes used, cleansed and sterilized.

I have mentioned, under the head of Bacteriological Work, that Cruikshank's Bacillus was added to the Applied Water on July 27, and August 17, and on October 11 and 12, at the rate of more than one million (1,000,000) per c. c., and that only upon one occasion were any traces of the same discovered, namely:—three in the sample of Filtered Water collected on July 27.

I consider the average efficiency of the Experimental Morison Mechanical Filter for removing water bacteria, from October 17 to November 11, 1893, and from January 24 to 30, 1894, at which time the filter was apparently free from growths of bacteria, as satisfactory as is generally obtained by Natural Filtration, with the best constructed sand filter-beds, at the rate of from 2,000,000 to 3,000,000 gallons per Acre per 24 hours.

It is stated in Senate Document, No. 4, of the State of Massachusetts, for 1894, that Bacteriological samples that were collected from October 13 to December 15, 1893, from the "Effluent" of the large filter that has recently been built at Lawrence, Mass., and that was put in operation September 20, 1893, and which was an outcome of investigations of the State Board of Health of Massachusetts, have given an average result of bacteria removed of 98.16 per cent.

In regard to the removal of Bacillus Prodigiosus, I consider the working of the Morison Mechanical Filter very satisfactory in this respect, with the exception of the run that was made on December 29, when the result from the only sample that could be counted on account of liquefaction was 66.0 per cent.

I have previously stated that the average percentage of

efficiency of the filter for removing Bacillus Prodigiosus without including this 66.0 per cent. was 99.97 per cent.

It will be seen by table No. 19, that the number of bacilli in the Applied Water on December 29, was 95,500. This of course is a very excessive number as well as the number which appeared in the Applied Water on a number of other days, as the average number of water bacteria that have been found in the samples of Applied Water since October 17, has been about 5,400 per c. c.

I have made a careful study of the results of all of the runs that have been made by the Morison Mechanical Filter, including those runs, given in table No. 1, in the first part of the report, making in all the results of one hundred and fifty-six (156) runs, and I have not been able to discover anything which would lead me to think that an occurrence similar to that of December 29, had taken place at any other time, and assuming that it might happen once in one hundred and fifty-six (156) times, as was the case during our experiments, the percentage of likelihood of its taking place would be $\frac{64}{100}$ of 1 per cent.

The average efficiency of the filter during the four runs when Cruikshank's Bacillus was used, that I have previously mentioned, was, for removing these bacilli, of course, 100 per cent.

Table No. 20, shows, from an average of three analyses, a reduction of *Albuminoid Ammonia by Filtration, of 70.0 per cent.*, and a reduction of *Ready-formed Ammonia of 91.0 per cent.*

It has been computed from figures given in tables, on pages from 469 to 489, of the Report of the State Board of Health of Massachusetts for the year 1892, that 58 per cent. of the Albuminoid Ammonia and 88 per cent. of the "Free Ammonia" was removed from the Applied Water by Natural Filtration at Lawrence, Mass., from June to November, with the experimental filters which are mentioned in table No. 22 of this report as having beds about five (5) feet deep.

The following table gives the proportion of Sulphur trioxide (SO_3), that was found by analysis in a sample of Pawtuxet River Water, before and after being treated with a solution of Sulphate of Alumina. The results of analyses of other waters are also given for comparison :—

From Report of Professor Thomas M. Drown, which is appended to this report.	Sulphur trioxide (SO_3) in Parts per 100,000.
Pawtuxet River Water, after being filtered through a double thickness of Swedish paper	0.5357
This filtered water after being treated with a solution of Sulphate of Alumina, in the proportion of one-half (½) grain of the sulphate to the gallon, and agitated for about one minute............ ...	0.8928
(The very slight flocculent precipitate produced was filtered off through a double thickness of filter-paper).	
Increase due to the addition of Alumina.........	0.3571
From Report of the State Board of Health of Massachusetts for the year 1892. (Page 346).	
Normal Ground Water.	
Mansfield, well..................................	0.131
Large Population in Drainage Area.	
Stoughton, well..............................	2.039
Everett, spring..............................	1.536
Malden, tubular wells......................	4.184
Imperfect Natural Filtration from Unpolluted Reservoir.	
Wayland, reservoir..............................	0.196
Wayland, filter-gallery..........................	0.136
Wells Containing Considerable Iron in Solution as the Result of Organic Matter and Iron in the Ground.	
Westborough, insane hospital, tubular wells.......	0.512
Reading, filter-gallery........................	5.960
Bradford, Well No. 7...........................	0.707
Bradford, Well No. 12..........................	0.374

From Report of the State Board of Health of Massachusetts for the year 1892. (Page 346). *Continued.*	Sulphur trioxide (SO_3) in Parts per 100,000.
Wells Near the Sea.	
Marblehead Water Company, Swampscott, large wells and tubular wells.	2.980
Marblehead, town supply, large well and tubular wells.	3.060
From Transactions American Society of Civil Engineers, Vol. XXX, 1893.	
Well, near Dresden, Germany, the water of which, it is stated, is of exceptional purity	3.480
A spring, near Providence, R. I., (the water from which is extensively sold in Providence).	1.540

The Treasurer of the Providence Dyeing, Bleaching, and Calendering Company, of Providence, R. I., which has a Morison Mechanical Filter Plant in operation and has used Filtered Water in six tubular boilers since August 1, 1893, has furnished the following information: The estimated quantity of Alumina added to the Applied Water was at the rate of about one-half (½) grain per gallon. The six boilers of the works were inspected June 1, 1893, before the use of Filtered Water was commenced, and on December 5, 1893, after Filtered Water had been used in the boilers about four months. At this last inspection, December 5, that was made to ascertain, if possible, if the Filtered Water, that had been treated by Alumina, had affected the the boilers in any way, no scale or corrosion was detected in the pipes or boilers that could be attributed to the use of Alumina.

Sixth.—The percentage which the Color of the water would be reduced by Filtration, viz.:—

The Color of the Applied and Filtered Water was determined from standard samples ranging from 1 to 10, that were prepared by Professor J. H. Appleton, of Brown University.

Table No. 21, gives the average Color results that were obtained during the day and night as well as the percentage of Color that was removed by Filtration.

The first Color sample of Filtered Water that was collected after the filter was started, was collected one (1) minute after water commenced to flow, then five samples were collected every five (5) minutes for one-half hour, and after that every hour during the day and night.

The Applied Water samples were collected every hour during the day, and four times during the night, namely:—from 5 P.M. to 8 A.M.

The averages given in table No. 21, apply only to runs during which Basic Sulphate of Alumina was used, and as will be seen by the table, the Color observations were not commenced until the experiments had been in progress some time.

The average percentage of Color removed from June 28 to October 26, 1893, as will be seen by the table, was 81.0 per cent., and the average percentage of Color removed from October 26, 1893, to January 30, 1894, was during the day 78.0 per cent., and during the night 66.0 per cent.

The difference in the percentage of Color between the day and night samples may have been largely due to the difference in the length of time that elapsed between the time that the samples of water were collected and the time that they were compared. The samples that were collected during the day being compared very soon after they were collected, while the samples that were collected during the night, which were not compared until morning, were kept from 13 to 5 hours. A difference in the constituency of the Applied Water may also have exerted an influence, as the majority, if not the whole, of the manufactories located upon the banks of the river, that supplied the settling basin from which the Applied Water was taken, were shut down during the night, and the quantity of water flowing in the river was generally less during the night than it was during the day.

There was always a very perceivable improvement in the appearance of the water after it had passed through the filter, and I consider that the efficiency of the filter in reducing the Color of the Applied Water was very satisfactory.

From a table on page 468 of the Report of the State Board of Health of Massachusetts for the year 1892, showing the efficiency of the experimental filters constructed in 1892, at Lawrence, Mass., for the removal of Color, by Natural Filtration, with filter-beds five (5) feet deep, after different lengths of service, it has been computed, that after 100,000,000 gallons of water had been filtered, 60 per cent. of the Color of the Applied Water was removed, and after 400,000,000 gallons had been filtered, 45 per cent.

Seventh.—The washing of the filter-bed, viz.:—

Our experiments have shown, I think, that the filter-bed could be very thoroughly washed by the aid of the mechanical rake or agitator.

Eighth.—The time which would be required for washing the filter-bed, viz.:—

The average time required for washing the filter-bed was about eleven (11) minutes.

Ninth.—The quantity of water which would be required to wash the filter-bed, viz.:—

The quantity of water required to wash the filter-bed, based on a run equivalent to a rise of four (4) feet of water in the filter after the full capacity had been reached, averaged about 4.9 per cent. of the quantity of water that was filtered during each run.

Tenth.—The quantity of water which it would be necessary to run to waste after washing the filter-bed, viz.:—

The quantity of water required for this purpose was based on the water running to waste Thirty (30) Minutes, for the reason that I have previously mentioned, and averaged about 2.9 per cent. of the quantity of water that was filtered during each run.

Eleventh.—The length of time which the filter would run after starting, before it would be necessary to shut down and wash the filter-bed, on account of the water gradually rising to its prescribed limit in the filter, on account of the filter-bed becoming gradually clogged up, viz.:—

During the runs when one-half (½) grain of Alumina and "Free Flow" were used, the average length of time that it took for the water to rise in the filter four (4) feet above the point where it stood when the filter first commenced to discharge, at its full capacity, (128,000,000), was, since October 17, 1893, 16 hours and 43 minutes. The length of time ranged from 13 hours and 22 minutes, to 19 hours and 44 minutes. The average length of time required from April 26, 1893, to January 30, 1894, was 16 hours and 53 minutes, ranging from 6 hours and 27 minutes, to 26 hours and 53 minutes.

During the runs when three-fourths ($\frac{3}{4}$) of a grain of Sulphate of Alumina and "Free Flow" were used, since October 17, the average length of time was 15 hours and 2 minutes, ranging from 13 hours and 28 minutes, to 17 hours and 37 minutes.

Our experiments have shown that the length of time which the filter will run, does not increase in the same proportion when the water is above four (4) feet that it does when the water is below four (4) feet. The proportion from three (3) to seven (7) feet is about three-fourths ($\frac{3}{4}$) of what it is below four (4) feet, for example, considering this proportion, the average run of the filter when one-half ($\frac{1}{2}$) grain of Sulphate of Alumina and "Free Flow" were being used, for a height of six (6) feet would be approximately, $(\frac{16 \text{ h. } 43 \text{ m.}}{6 \text{ ft.} - 4 \text{ ft.}} \times .75) + 16$ h. 43 m. = 22 h. 59 m.

The condition of the water in the River, from which the Applied Water was indirectly taken, since the experiments were first commenced, has been quite varied; when it has not been in its normal condition, it has sometimes been at flood height, sometimes remarkably low, sometimes containing considerable suspended matter that has been washed into it by heavy rains from the surface of the ground, sometimes containing considerable suspended matter such as leaves, etc., that have fallen from the trees and bushes located along its banks, and sometimes it has been remarkably free from suspended matter.

The length of any of the runs of the filter did not materially change on account of the different conditions of the water in the river, with the exception that, occasionally during the summer months, after heavy rains, the length of time was reduced about one-half ($\frac{1}{2}$).

Twelfth.—The effective stability of the quartz and the supplementary precipitate bed:—whether it could be depended upon to do its work thoroughly during the time that the filter was in operation or whether at times it would be liable to crack or break, or have its efficiency reduced in any manner, viz.:—

The number of observations that had been taken daily previous to September 6, were increased on this date for the purpose of investigating the stability of the filter-bed, and the number of bacterial samples collected daily, were also increased on November 23, for the same reason.

The height of water in the filter was observed every half-hour during the day and every hour during the night.

The quantity of Applied Water flowing into the filter and the color of the Filtered Water were observed every hour, day and night.

The results of ninety-two (92) runs of the filter, that were made during the time that these observations were taken, were plotted in the manner shown on Diagram No. 1.

Each result was carefully examined. If the line representing the half-hourly and hourly heights of water in the filter showed a gradual increase in proportion to the time, without any sudden break, it was considered an indication that the supplementary precipitate bed had not cracked or broken and that the quartz-bed was in good condition. If the line representing the half-hourly and hourly heights of water was abruptly broken, and the line representing the hourly quantity of Applied Water was also abruptly broken at the same time, it was considered an indication that the falling off of the elevation of the water in the filter was due to the decrease of the hourly quantity of the Applied Water, which was generally owing to some impediment having got into the meter or valve on the supply-pipe, such as a small eel or some other obstruction. The color line on the Diagram being uniform or nearly so, corroborated this supposition. If there had been a sudden break in the line representing the half-hourly and hourly heights of the water in the filter, and if the line representing the hourly quantity of Applied Water had not shown any change, and the line representing the color of the Filtered Water had suddenly shown an increase at the same time, it would have been considered as an indication that the supplementary bed had broken or that the quartz-bed had given way. There was not anything of this kind discovered, however, and a careful study of the plotted results, as well as of the hourly bacterial results, indicated that there had not any unfavorable changes taken

Diagram No. 1

Showing Elevation of the water in the filter, Hourly readings of the Meter on the supply pipe, and the Color of the Applied and Filtered water.

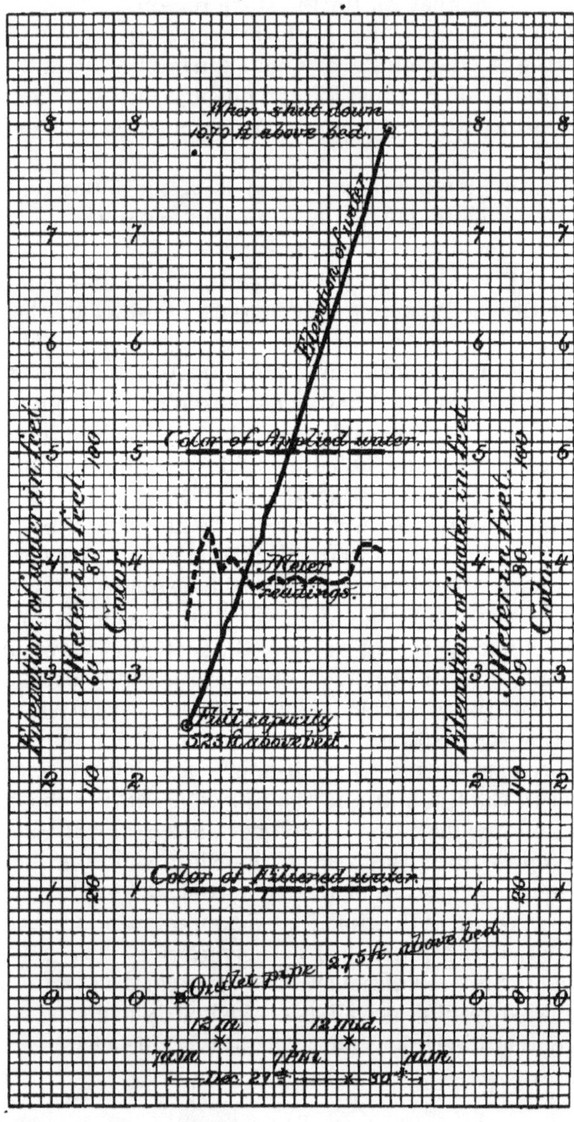

place in the filter-bed due to cracks or breaks, etc., with the exception, possibly, of what is shown by the bacterial results of December 29, which I have previously mentioned in detail.

The working head of the filter (the elevation of the water above the point where it stood at the time the filter first began to discharge at its full capacity), ranged, when it was shut down at the end of the different runs, from three and five-tenths (3.5) to six (6.0) feet. The bacterial and other results obtained were equally as good with the head at six (6.0) feet as they were with the head at three and five-tenths (3.5) feet.

Three samples of the quartz filter-bed, that were taken at different depths on August 18, at the end of a run of 23 hours and 19 minutes, after being treated in the manner that such samples are generally treated and cultivated two days, gave the following bacterial results:—

Approximate depth of Sample below the top of the bed, in Inches.	Number of Bacteria in One Gramme of Quartz. (Dry Weight).
¾	364,000
2	26,054
5 to 7	2,950

The total number of bacteria that flowed into the filter with the Applied Water, during the run of 23 hours and 19 minutes, was about 23,710,000,000, assuming that the number per c. c. in the Applied Water was the same as was found in the sample of Applied Water that was collected at 12. M.

I have estimated, approximately, by the aid of the above table, that a layer of the quartz filter-bed five and one-quarter (5¼) inches deep, extending from a point three-fourths (¾) of an inch below the top of the bed to a point six (6) below the top, contained at the end of the run about twenty-four (24) per cent. of the total number of bacteria that had flowed into the filter in the Applied Water during the run, and that the remaining lower portion of the bed contained about three (3) per cent. of the total number. The other seventy-three (73) per cent., with the exception of about one (1) per cent. which may have gone through the filter-bed into the Filtered Water, was probably lodged upon and in the supplementary gelatinous precipitate bed upon the top of the quartz-bed and in the upper three-fourths (¾) inch layer of the quartz-bed.

A sample of the Effluent Wash Water, collected at the end of a run of 24 hours, on August 4, immediately after the "agitator" was started, at the time that the water was removing the supplementary gelatinous precipitate bed and when it was the blackest and dirtiest, indicated that a one and one-fourth (1¼) minutes flow of this Effluent Wash Water, provided that the total number of bacteria per c. c. continued to be the same as when the sample was taken, would contain a number of bacteria equal to the total number that flowed into the filter in the Applied Water during the run, assuming that the number in the Applied Water per c. c. was the same as was found in the 12 M. sample. The indicated efficiency of the filter during this run ending August 4, was about 100 per cent.

Thirteenth.—The loss of head due to the water flowing through the filter-bed, screens, and outlet-pipe, viz.:—

The average loss of head due to the passage of the water through the supplementary precipitate bed, the quartz-bed, screens, etc., at the moment when the filter had reached a capacity of 128,000,000 gallons per Acre per 24 hours, was 2.44 feet.

From a mechanical standpoint, the working of the Experimental Morison Mechanical Filter, throughout the experiments, was very satisfactory.

I have estimated $5.69 as the cost, per 1,000,000 gallons, of operating a Morison Mechanical Filter Plant (having an effective capacity of 15,000,000 gallons per 24 hours), throughout the year.

Removal of Water Bacteria by Subsidence and Flow.

During our filtration experiments, at different times in April, May, July, August and September, 23 samples of the Pawtuxet River Water were taken at the same time (about 12 M.), that the regular samples of Applied Water were taken. It was found from these samples, after a cultivation of about 48 hours, that 40 per cent. of the bacteria was removed from the River Water before it reached the Applied Water tap. The River Water travelled in its course to the Applied Water tap, at the rate of 9,000,000 gallons per 24 hours, through a thirty-six (36) inch pipe one hundred and fifty (150) feet long, a settling basin having a capacity of 5,850,000 gallons, a thirty (30) inch pipe one hundred and seventy-three (173) feet long, a three-sixteenths ($\frac{3}{16}$) inch mesh screen, a pump-well having a capacity of 18,000 gallons, four (4) pumps, a thirty (30) inch pipe two hundred and seventy (270) feet long, and at the rate of 14,400 gallons per 24 hours, through a two (2) inch pipe one hundred and fifty (150) feet long. Samples of the River Water and Applied Water, were also taken on June 1, hourly from 8 A. M. to 3 P. M. inclusive, for the same purpose mentioned above, and under the same conditions, and it was found that the average reduction of the River Water bacteria on this day, in its course to the Applied Water tap, was thirty-nine (39) per cent. As there were one or more springs found in the bottom of the settling basin during its construction, I should say, taking into consideration some measurements that I have made, that it is possible, though hardly probable, as the surface of the water in the settling basin was kept practically at the same height as the surface of the water in the river, that 5 per cent. of the water flowing into the basin, per 24 hours, may have flowed in through subterranean sources.

FINIS.

I have consulted in regard to the Bacteriological Work, nearly every day, while the experimental filtration work was in progress, with Dr. C. V. Chapin, Superintendent of Health, with the exception of about six weeks, when he was confined to his house by illness or was out of town. I am much indebted to Mr. George W. Fuller, in charge of the Experimental Station of the Massachusetts State Board of Health at Lawrence, Mass., for information that he has furnished me from time to time, in regard to the methods that are followed in the cultivation of bacteria at that station. I am also indebted to Professor J. H. Appleton of Brown University, Professor Charles A. Doremus of the College of the City of New York, and Professor T. M. Drown and Mrs. E. H. Richards of the Massachusetts Institute of Technology, for advice relative to some of the chemical problems that were encountered in the course of the work, and to Dr. G. T. Swarts, Medical Inspector, Professor H. C.

Ernst of the Harvard University Medical School, Professor E. K. Dunham of the Carnegie Laboratory, and Professor T. M. Prudden of Columbia College, for advice in connection with the bacteriological work.

Attention is called to the appendix of this report containing a report of Professor T. M. Drown, and a number of letters, references and tables, germane to the subject of the chemical purification of water.

Mr. James A. McKenna of the City Engineer's Department, had charge, during the experimental filtration work, of operating the different experimental filters and collecting the samples of water at Pettaconset Pumping Station.

Respectfully submitted,

EDMUND B. WESTON,

Assistant Engineer in charge of Water Department.

Appendix.

To the Report of Edmund B. Weston, C. E., upon the Results Obtained with the Experimental Filters at the Pettaconset Pumping Station, of the Providence Water Works, containing:—

Report of Professor Thomas M. Drown, upon the analysis of a sample of Pawtuxet River Water, before and after adding one-half (½) grain of Sulphate of Alumina.

Letters from the Hartford Steam Boiler Inspection and Insurance Co.

Letter from Dr. C. V. Chapin, giving information relative to "mechanical filtration," obtained by personal inquiries, inspection and correspondence.

Extracts from two papers, published in the Transactions of the American Society of Civil Engineers, relative to the use of Alum in the purification of water.

Extracts from an article, published in the Chemical News, relative to experiments with Alum Baking Powders, etc., etc.

Letter from Professor C. A. Doremus, relative to the action of purified waters upon boiler scale.

Letter from the Treasurer of the Providence Dyeing, Bleaching and Calendering Co., relative to experience with filtered water in boilers and wrought iron pipes.

Table giving the grains per gallon of "Alumina" and "Sulphuric Acid," which are contained in 146 Mineral Springs of the United States.

Table giving the grains per gallon of "Alumina" and "Sulphuric Acid," which are contained in some Natural Waters in Massachusetts.

Table showing the Number of Times, during 15 Winters, from 1880-81 to 1894-95 inclusive, that periods occurred (Number of days without intermission), of 1 day and More, when the Daily Mean Temperature was 32° and Less, at Providence, R. I.

Table showing the Normal Mean January Temperature in Degrees F., of a number of European cities and at Providence, R. I. Also, the Normal Mean Temperature at Providence, R. I. for December and February.

Table showing the Number of Times and the Number of Days in each Period during each Winter from 1879-80 to 1886-87 and from 1888-89 to 1892-93 that the Hope Reservoir of the Providence Water Works was Frozen Over. Also, the Date that the Reservoir was Frozen Over the First Time during each Winter and the Last Time that Ice was visible, and the Total Number of Days that the Daily Mean Temperature was 32° and Less.

Cost of Mechanical and Natural Filtration.

Report of Professor T. M. Drown.

MR. J. HERBERT SHEDD,

City Engineer,

Providence, Rhode Island.

DEAR SIR:

At your request I have made experiments to determine the effect of the addition of one-half grain of sulphate of alumina to Pawtuxet River Water, and I give you herewith the results of the investigation.

The sulphate of alumina sent me had the following composition:

	Per cent.	One-half grain contains in grains.
Insoluble residue.	0.52	0.0026
Alumina (Al_2O_3).	15.78	0.0789
Sulphuric acid (SO_3)	36.79	0.1840
Water (by difference)..	46.91	0.2345
	100.00	0.5000

The Pawtuxet River water contained, after being filtered through a double thickness of Swedish paper, the following mineral substances:

	Parts per 100,000.	Grains per gallon.
Silica..	.3833	.2238
Ferric oxide (Fe_2O_3).	.0567	.0331
Alumina (Al_2O_3).	.0500	.0292
Lime (CaO).	.4367	.2550
Sulphuric acid (SO_3).	.5357	.3129

This filtered water was treated with a solution of the sulphate of alumina in the proportion of half a grain of the sulphate to the gallon, and agitated for about one minute. The very slight flocculent precipitate produced was filtered off through a double thickness of filter paper. The filtrate had the following composition:

	Parts per 100,000.	Grains per gallon.
Silica (SiO_2).	.3866	.2257
Ferric oxide (Fe_2O_3)..	.0167	.0098
Alumina (Al_2O_3).	.1000	.0584
Lime (CaO).	.4433	.2588
Sulphuric acid (SO_3)	.8928	.5214

On comparing these two analyses it will be seen that silica and lime are practically the same in both, that the iron is notably less in the water which has been treated by the sulphate of alumina, and that the alumina and sulphuric acid are both considerably higher. The amounts are as follows:

	Grains per gallon.
Ferric oxide, decrease..........	.0233
Alumina, increase.	.0292
Sulphuric acid, increase....	.2085

The cause of the decrease of the amount of iron is the partial removal of the coloring matter of the water by the precipitated alumina. The coloring matter of brown waters is composed of organic matter and iron, and when the organic matter is removed by the mordanting power of the alumina, the iron oxide is precipitated, having nothing to keep it in solution. This is a fact which I have frequently observed in treating brown waters with alumina.

The amount of sulphuric acid in the filtered water, after treatment with sulphate of alumina, should be the amount in the sulphate in addition to that in the original river water. In the actual determinations made, the amount of sulphuric acid is slightly higher than the calculated amount, but not more than may be attributed to the limits of accuracy of the analytical processes.

	Grains per gallon.
Sulphuric acid after treatment with Sulphate of Alumina.	0.5214
Ditto, in original water..........................	0.3129
Increase due to the Sulphate of Alumina.	0.2085
Amount of Sulphuric acid in one-half grain of Sulphate of Alumina..........	0.1840

The increase in the amount of alumina in the water after treatment with sulphate of alumina and filtering, is 0.0292 grain per gallon. The amount of alumina added in one-half grain of the sulphate is 0.0789 grain. This would indicate that somewhat over one-half of the sulphate was decomposed, and its alumina precipitated; the remaining portion passing into solution. Thus:

	Grains.
Amount of alumina in one-half grain of sulphate.	0.0789
Amount of alumina in one gallon of filtered river water...	0.0292
Total............	0.1081

Amount of alumina in one gallon of water after treatment with sulphate and filtering....	0.0584
Amount of alumina precipitated..........	0.0497
Increase of alumina in water after treatment with sulphate and filtering......	0.0292

I do not attach any greater importance to these determinations of alumina than to show what took place in this single experiment. I am inclined to think from similar experiments which I have made from time to time, that the amount of alumina precipitated when a weakly-alkaline, natural surface-water is treated with a minute amount of sulphate of alumina is dependent on time, on the amount of agitation, and also on the degree of alkalinity of the water, which may vary from time to time. I think also that the precipitated alumina sometimes is redissolved in the water, in part at least, on long standing, particularly when the water has not been completely decolorized by the alumina. It is not safe to reason *à priori* from our knowledge of what takes place in moderately dilute saline solutions to what will take place in excessively dilute saline solutions in water containing considerable organic matter.

Under the conditions employed in this experiment the results show that there is more alumina in the water after treatment with sulphate of alumina and filtration than there was in the natural water. This increase is 0.0292 grain per gallon, which happens to be the amount present in the original water.

As to the condition in which the alumina is combined in the water, whether in its original form as sulphate, or in some other combination, it is, I think, impossible to say. The usual disposal of the acid and basic radicals among each other, as the result of an analysis of the solid residue of evaporation of a water, is largely speculative, and does not throw much light on the condition of these substances when in solution.

The increase of sulphuric acid in the water after treatment with sulphate of alumina might be a positive disadvantage when used for boilers, if the water contained sufficient lime to combine with the increased amount. In the case of the Pawtuxet River water the amount of lime is only slightly in excess of that required by the sulphuric acid naturally present in the water, so that the increase of sulphuric acid after treatment with sulphate of alumina can only form a very small amount of additional boiler scale.

From the analyses I calculate that it would require the evaporation of 5,000 gallons of water, after treatment with sulphate of alumina, to give one ounce of additional boiler scale of sulphate of lime. A boiler evaporating 10,000 gallons in 24 hours would thus accumulate a scale of one pound of sulphate of lime in eight days, over and above the amount produced by the use of untreated Pawtuxet River water.

Finally I have determined the effect on the color of the water by treatment with sulphate of alumina. I found that the use of one-half grain of the sulphate reduced the color from 0.45 to 0.18. One grain to the gallon rendered the water practically colorless, namely, 0.02, and two grains reduced the color to 0.01. These figures refer to the standards of color used by the Massachusetts State Board of Health.

THOMAS M. DROWN.

Massachusetts Institute of Technology.

Boston, July 12th, 1893.

[In the above report Professor Drown uses the term "sulphuric acid" for SO_3. He probably makes use of the term in a popular sense, as in reality SO_3 is "sulphuric anhydride" or "sulphur trioxide." The correct symbols for sulphuric acid are H_2SO_4.]

E. B. W.

Copied Letter.

HARTFORD STEAM BOILER INSPECTION AND INSURANCE CO.

HARTFORD, CONN., April 15, 1893.

Chemical Department.

Analysis No. ...

J. M. ALLEN, ESQ., President,

Hartford Steam Boiler Inspection and Insurance Co.,
Hartford, Conn.

DEAR SIR:

Your note, with letter of Chas. V. Chapin of Providence, to you, making inquiry concerning the effect of alum treated waters on steam boilers, has just been received.

Alum, being a sulphate with an acid reaction, is much more injurious to a boiler than any of the salts ordinary found in waters that are inclined to cause corrosion: and the presence of a small excess of unchanged alum in a water will speedily cause quite serious corrosions, particularly if confidence in the improved quality of the water causes the boiler user to blow off and so change the water much less frequently than before, so that the alum solution becomes more and more concentrated.

I have met quite a number of cases where boilers are rapidly rusted by alum treated waters, even in regions of "hard" lime bearing waters, where the alum treatment is most efficacious.

In order to completely purify a water by any of the alum processes, it would appear that the alum should be very slightly in excess to insure perfect results, and it seems probable that in New England waters which often, and sometimes with quite rapid variation, contain but the slightest amount of Lime Carbonate, and of such organic matter on which an alum coagulation largely depends, the excess of alum without great care might be considerable.

However, corrosion from the use of alum is not so general as might be supposed, and I attribute this to be because either the alum is generally used in insufficient amount to entirely clear the water, or that the water either naturally contains soluble alkalies, or the rather general use at present of Soda Ash and other alkaline solvents in boilers prevents injurious action of any alum in the water.

Lime Sulphate, a product of the use of the alum process, is so appreciably soluble in water as to cause trouble from formation of hard and intractable scale, second only to the alum itself. I should not consider it safe, where alum purification is employed, to run a boiler without a sufficiency of Soda Ash or similar boiler compound, to keep the water slightly alkaline, ½ to 1 lb. per 1,000 gallons would be sufficient in New England, and would also decompose the Lime Sulphate to the manageable Carbonate. If a reddish powdery deposit (iron rust from alum corrosion) is noted, more Soda Ash should be used.

Respectfully,

GEORGE H. SEYMS, CHEMIST, (Signed)

Hartford Steam Boiler Inspection and Insurance Co.

Copied Letter.

HARTFORD STEAM BOILER INSPECTION AND INSURANCE CO.
SOUTHEASTERN DEPARTMENT,
B. F. JOHNSON, Chief Inspector,
Atlanta, Georgia, June 27, 1893.

CHAS. V. CHAPIN,
Supt. Board of Health, Providence, R. I.

DEAR SIR:

Your letter, of June 19th at hand, asking for information regarding the filtering system used by this city.

I have most of the boilers, in this city, insured and under my charge. Thus far, have been unable to find any bad effect on the iron or steel, from the alum used in purifying the water. There being no corrosion or cutting, that I considered, came from the alum.

I have been inspector for the Hartford company, nearly six years, and have had the boilers under my charge ever since.

We have only had two batteries of boilers, in this city, that have ever given us any trouble in the way of corrosion or pitting. These batteries of boilers were located in different parts or the city and were on the end of pipe lines. Very little water was used from the street mains, except, what was used at these plants and most of the corrosion and pitting was due to rust and sediment, that settled in the mains. When valves were opened up, it was carried on, through, into the boilers.

Some time ago, one of the locomotives used for shifting cars, in this city, gave way in the fire-box.

There was a statement in the papers, that it was caused by the alum used for filtering the water that cut away the stay-bolts and caused the fire-box to let go. This statement was not true. The stay-bolts were broken off, by the jar and long service. Scarcely any cutting could be found on the stay-bolts.

Respectfully yours,

B. F. JOHNSON.

Copied Letter.

HARTFORD STEAM BOILER INSPECTION AND INSURANCE CO.

Hartford, Conn., July 17, 1893.

JOHN A. COLEMAN, ESQ.,

41 Wilcox Bldg., Providence, R. I.

DEAR SIR:

The question in regard to the percentage of alum in a gallon of water that would be injurious to a boiler, was referred to the undersigned by our 2d Vice President Mr. F. B. Allen. I immediately called the attention of our Chemist to the subject and he replied that in his judgment one-half grain of alum in a gallon of water would not be injurious and would have no appreciable effect on the boiler. This report we supposed you had received, but it seems you had not. Large quantities of alum in water, particularly water that has lime in solution is not a good thing, as a resultant of the use of alum under such circumstances would be an increased amount of lime sulphate in the water, which produces a hard scale, so troublesome in some parts of New England; but one-half grain, per U. S. gallon in our judgment, would do no harm.

Truly yours,

J. M. ALLEN, PRESIDENT.

Copied Letter.

HEALTH DEPARTMENT, OFFICE SUPERINTENDENT OF HEALTH,

City Hall, Providence, June 11, 1894.

DEAR MR. WESTON:

In 1889 I visited Long Branch and examined a filter there in operation in which alum was used as a coagulent. There was no taste of alum in the water and I questioned a number of people that I met at the hotel and other places as to the presence of alum in the water and all claimed that they had never noticed it. I was referred for further information to Dr. Hunt whose statement you will find on page 43 of my report for 1889.*

I at that time wrote to the health officers of places using water that had been treated by alum, and received answers which you will find in the Report named. Alum has also been used for some years at Newport, though somewhat irregularly, and I have repeatedly questioned Newport physicians in regard to its effect upon the users of the water, and they all say that they have never noticed any deleterious effects. During 1893 I wrote to the health officers of some forty or fifty towns using alum-treated water, and received replies from the following places: Atlanta, Ga.; Bordentown, N. J.; Chattanooga, Tenn.; Elgin, Ill.; Exeter, N. H.; Independence, Kan.; Lakewood, N. J.; Little Rock, Ark.; Macon, Ga.; Mt. Clemens, Mich.; Mt. Pleasant, Ia.; New Orleans, La.; Ottumwa, Ia.; Owego, N. Y.; Porterville, Cal.; Richfield Springs, N. Y.; St. Thomas, Ont.; Somerville, N. J.; Sidney, O.; Trenton, Mo.; Tuckhannock, Pa.; Waterloo, Ia. In no case were any ill effects attributed to the use of water which had been treated with alum and filtered. In many instances it was stated that the effluent water had been tested by competent chemists and no alum found. In one or two instances it was stated that alum had occasionally been found in the water, but the amount was small and no bad results had been noticed. I made particular inquiries at Lakewood, N. J., through a friend who was passing the winter there. He saw the best physicians and examined into the subject quite carefully, and stated it was the unanimous opinion that the filtration of water by the Hyatt process, in which alum is used as a coagulent, resulted only in its improvement.

During June of last year I visited Chattanooga and Atlanta, and had personal interviews with the health officers and other persons in both places, and their opinion as to the harmlessness of using the water was the same as that expressed in 1889. During the visit I paid particular attention to the effect of the filtered water upon boilers. At Chattanooga I visited a planing mill, an ice factory, two flour mills, and the power plant of the street railway company; and all these users of steam were emphatic in their statements that the use of the filtered water did not produce either scale or corrosion and were all highly pleased with the process of filtering because the sediment was so much less. At Atlanta, I visited, among other places, two of the largest cotton mills, a paper mill, two ice factories, and the power plant for the street railway. All the parties visited made statements similar to those made by the users of steam at Chattanooga. I saw also the boiler-makers there who do most of the repairing and they also said that it was their opinion that filtration improved the water and that the use of alum was never the cause of corrosion. The letter which I received from the inspector of the Hartford Steam Boiler Inspection Company in Atlanta, I have already given to you.

Yours truly,

CHARLES V. CHAPIN.

* The following letters were copied from Doctor Chapin's Report for 1889:

CHILLICOTHE, MO., Nov. 2, 1889.

DR. CHAPIN, *Sup't. Health.*

DEAR SIR:

Yours of 29th received. In reply I will say that I have called upon all of the physicians of our city to ascertain if possible whether or no their attention had been called to the matter of alum in water, and as to whether there had been any complaint as to the purity of the water furnished our city. The prevailing opinion of our physicians and people is that the water furnished by the water works is far superior to any well water that we have in the city. As for myself and family, we use the water, and I have found no reason for complaint. We have made no analysis of the water, but as my attention has been called to the matter by your letter, I shall attend to the matter and give it a fair test. Undoubtedly, if alum is present in the water, it will be injurious to the extent of amount present.

Yours truly,

S. M. BEEMAN, HEALTH OFFICER.

CHATTANOOGA, TENN., Nov. 1, 1889.

DR. CHARLES V. CHAPIN, *City of Providence.*

DEAR SIR:

Yours of the 29th ultimo to hand. In regard to the influence or effect of alum in our filters, I beg leave to say that they, the filters, have been in use in this city about two years. Prior to that time we took our water straight, without filtering. Since the water company have put in their filters I have noticed no increase in the amount of sickness or in death rate in our city, but, on the contrary, we are improving. Our water company are using the National filter, but I do not know to what extent they use alum.

Yours respectfully,

J. L. GASTON, M. D., PRESIDENT BOARD OF HEALTH.

99 E. MITCHELL STREET, ATLANTA, GA.,
Oct. 31, 1889.

CHARLES V. CHAPIN, M. D.,

Superintendent of Health, Providence, R. I.

MY DEAR SIR:

In answer to your inquiry of the 28th instant, I will state that the Hyatt system of filtration has been in use in this city for about two years. A small portion of alum—never exceeding one grain to the gallon—is automatically injected into the water just before it enters the filters. It serves the purpose of a coagulant, and it is claimed that no part of it passes out of the filters with the clarified water. Certainly no part appreciable to the senses remains. I have used the filtered water and no other in my family constantly, and have seen no evil effects from it. In a large general practice of medicine I have never heard any complaint or observed any disorder that I could ascribe to it. The water is of crystal-like clearness, is sparkling and palatable to a grateful degree.

Yours truly,

JAMES B. BAIRD.

LONG BRANCH, N. J., Feb. 15, 1890.

DR. CHARLES V. CHAPIN.

DEAR SIR:

I have not seen or heard of any deleterious or injurious effects from the use of alum in purifying city water.

Respectfully,

S. H. HUNT.

Extracts from two papers, relative to the use of Alum in the purification of water, published in the Transactions of the American Society of Civil Engineers.

Vol. XXX., 1893. EXPERIENCES HAD DURING THE LAST TWENTY YEARS WITH WATER WORKS HAVING AN UNDERGROUND SOURCE OF SUPPLY. By B. Salbach, Baurath at Dresden, Saxony, Germany.

The results found at the water works built by the author for the city of Groningen, in Holland, will be of general interest.

After heavy rain storms, and during spring freshets in the river, the water is colored brown by turf bogs situated further up stream and retains its yellowish-brownish color after filtration, although otherwise clear. To remove this color, a small quantity of alum is added to it by a small pump, while on its way to the settling basins. This addition of a saturated solution of alum amounts to about 1 to 10,000 or 1 to 20,000, of water. In the settling basins the alum is speedily distributed throughout the mass of water and greatly aids its clarification. The coloring matter contained in the water, principally iron, is precipitated, and the water, after settling from 12 to 14 hours and being filtered, becomes entirely clear and without color. A chemical examination of the filtered water shows that the alum disappears from the water after clarification, and that the sulphuric acid contained in the water is increased by a minute and hardly perceptible amount. A very important result was proved by the experiments of Professor Dr. von Calcar, of Groningen, viz., that during this operation, every trace of bacterial life had vanished.

The system has been tried at this place for 13 years and may, in other like cases, be of great service.

Vol. XXI., 1889. THE VICKSBURG SETTLING BASINS. By Clarence Delafield, M. Am. Soc. C. E.

Ordinarily, the water is now found clear and admirable for use, but at times, when the stain of the swamps is in solution, a small amount of either alum or perchloride of iron is found necessary to convert it into a form which can be removed by precipitation or filtration. A solution of alum is now used, which is introduced in absolute quantity from a tank connected by a pipe to the suction main of vertical pumps, thus intimately mixing it with the entire body of water, and presenting every portion to its chemical action. The result is absolutely clear water.

The percentage of alum used is about one grain to the gallon, and it is entirely inert and harmless, even if not doing its proper duty.

These works have now been in use several months, and the water delivered to consumers is giving perfect satisfaction.

Extracts from an article published in the Chemical News of December, 1888.

EXPERIMENTS UPON ALUM BAKING-POWDERS, AND THE EFFECTS UPON DIGESTION OF THE RESIDUES LEFT THEREFROM IN BREAD. By Professor J. W. Mallet, University of Virginia.

Experiments upon the Influence on Digestion of Moderate Doses of Aluminum Hydroxide and Aluminum Phosphate Swallowed shortly before or along with Food.

Having been interested by the results of a few experiments made in my own person a year or two ago on the apparent interference with digestion of these substances, I have tried a larger number of such experiments under more carefully noted conditions and with definite quantities of the materials used, in order to test directly the physiological effect of the residues from alum baking-powders, so far as this can be determined by their action in the case of a single person.

The experiments were made with intervals of three or four days between them; the food taken was of various kinds, but always simple and wholesome, and not likely of itself to produce disturbance of digestion; there was no pre-existing derangement of the digestive functions when any experiment was undertaken; as much care as possible was taken to avoid any mere fancying of expected symptoms, and to state with moderation what was actually experienced.

While on two or three occasions, particularly with the smallest doses used, there was no clearly observable effect, the general tenor of the experiments seemed to establish beyond doubt on my part the fact that the ingestion of the aluminum compounds used produced an inhibitory effect on gastric digestion, while in some cases, particularly with the larger doses, and on the whole rather with the hydroxide than the phosphate for equal weights of the two, the interference with the course of digestion was very notable. There was no gastric pain, nor were there any other symptoms of gastric or intestinal irritation, but simply the well-known oppressive sensations of indigestion properly so called, lasting for a longer or shorter time, but generally for at least two or three hours after the taking of food.

The quantity of aluminum hydroxide swallowed in each experiment varied from 10 to 50 grains, the average for all the experiments being about 28 grains. The quantity of aluminum phosphate used, ranged from 10 to 100 grains, the average being 45 grains. These doses were intentionally made larger than the quantities of the aluminum compounds in question derivable from such an amount of bread as would usually be eaten at a time if alum baking-powder in anything like usual proportion had been employed in making it. The object was to ascertain with what doses distinct effects were noticeable, and this seemed to be generally the case with any dose not less than 20 grains of the hydroxide or with not less than 30 or 40 grains of the phosphate. It may, of course, be reasonably supposed that a considerably less quantity than would be necessary to produce decided discomfort when once administered might prove objectionable and injurious if habitually taken as a part of the bread of each daily meal. With the proportion of alum in most of the baking-powders in use, with the allowance of two teaspoonfuls (counted as about 200 grains, though as much as 250 grains was found to be sometimes measured by a cook) of powder to a quart of flour, and assuming 35 or 40 per cent. of water in baked bread, a pound of bread would contain about 13 or 14 grains of aluminum hydroxide if alum alone were used in making the powder, or about 20 or 21 grains of aluminum phosphate if alum and calcium acid phosphate were used together, and all the aluminum were left in the bread as phosphate.

[As is given above, from 10 to 50 grains of Aluminum hydroxide, or hydrate, was swallowed in each experiment, and that no distinct effects were noticeable with doses less than 20 grains. Also, that from 13 to 14 are generally contained in the amount of baking-powder, containing alum, used with a quart of flour.

It is shown on page 43 that one-half (½) grain of Sulphate of Alumina, the average amount added per gallon to the water during the experiments with the Morison Mechanical Filter, contains only about 0.08 of a grain of Alumina ($Al_2 O_3$). The result of the addition of the one-half (½) grain of Sulphate of Alumina to the water, so far as Aluminum Compounds are concerned, was the formation of about 0.12 of a grain of Aluminum Hydroxide, or Hydrate ($Al_2 H_6 O_6$), which was precipitated upon the filter-bed, and retained within the filter, with the exception of a minute portion that came through the filter-bed with the water that was being filtered, which caused the discoloration produced by the Logwood and Acetic Acid test, in the Filtered Water.]—E. B. W.

Copied Letter.

CHEMICAL LABORATORY OF BELLEVUE HOSPITAL MEDICAL COLLEGE, EAST 26TH STREET.

NEW YORK, June 16th, 1894.

MR. E. B. WESTON.

DEAR SIR:

Water such as you have in Providence and we in N. Y., which holds organic matter in suspension and a small quantity of lime in solution, forms a scale at times, consisting largely of floating impurities baked on the boiler shell and partly cemented there with the lime. When all suspended matter is removed by filtration, no new scale forms, and the old disintegrates gradually. Engineers frequently use rain water for a time in boilers to loosen scale on this principle. There is scarcely enough lime in Providence water to form scale of itself, even though calcium sulphate is practically insoluble at 3 atmospheres steam pressure. Below this what calcium sulphate there is in Providence water after filtration should remain dissolved, since its coefficient of solubility is such that unless the water were greatly concentrated by evaporation the saturation point would not be reached.

Even under high pressures, it would require the evaporation of many thousand gallons of water for even a thin coating of calcium sulphate to appear in a boiler.

Yours very truly,

CHARLES A. DOREMUS.

Copied Letter.

PROVIDENCE DYEING, BLEACHING AND CALENDERING CO.
(Founded 1814.)

P. O. Box, 1131.
Telephone, 1708.

52 VALLEY STREET, PROVIDENCE, July 18, 1894.

MR. E. B. WESTON.

DEAR SIR:

Our battery of six boilers was examined on Dec. 3d, 1893, by F. D. Terry, Inspector for the Hartford Steam Boiler Insurance Co. This was six months after we commenced to feed the boilers with filtered water that had Sulphate of Alumina added to it at the average rate of ½ gr. per gallon. There was nothing discovered during this examination of the boilers which indicated any injurious effects from the use of the filtered water.

When we first began to use the filtered water, a scale or deposit which our wrought iron pipes contained was acted upon and gradually removed by the purified water flowing through them. In consequence of this the water was at times very dirty until the scale was entirely removed, which took, with our somewhat irregular use of different pipes, some two weeks or more. Since then we have had no trouble with the old pipes, and we have never had any trouble with the new pipe.

Yours truly,

JOHN P. FARNSWORTH, TREAS.

The boilers referred to above were examined by the City Inspector on the 30th of May last, and were found in practically the same state as reported above.

TABLE A.

The following table gives the grains per gallon of Alumina or Aluminum Oxide and Sulphuric Acid or Sulphuric Oxide which are contained in the waters of 146 Mineral Springs of the United States. The table was compiled from "Lists and Analyses of the Mineral Springs of the United States," published in Bulletin No. 32 of the United States Geological Survey. It embraces all of the results which are given in the bulletin relative to the constituents above mentioned, when quantities are given, in cases in which the springs are considered as "resorts" or their waters used commercially.

NAME OF SPRINGS AND STATE IN WHICH THEY ARE LOCATED.	REMARKS.	GRAINS PER GALLON.	
		Alumina.	Sulphuric Acid.
MAINE.			
Poland Silica Springs	Used commercially and as a resort.		
Star Spring		0.32	
Rosicrucian Spring	Do.	0.03	
Hartford Cold Spring	Do.	0.24	
NEW HAMPSHIRE.			
Birchdale Springs	Do.		
Concord Spring		0.12	
Unity Springs	Resort.		
Iron Spring		0.04	
VERMONT.			
Guilford Mineral Springs	Used commercially and as a resort.		0.69
Middletown Springs	Do.		
Spring No. 1		0.10	
Sheldon Spring	Do.		0.51

TABLE A.—CONTINUED.

NAME OF SPRINGS AND STATE IN WHICH THEY ARE LOCATED.	REMARKS.	GRAINS PER GALLON.	
		Alumina.	Sulphuric Acid.
NEW YORK.			
Ballston Spa Springs	Resort.		
Artesian Lithia Spring	Used commercially.	0.08	
Washington Lithia Well	Do.	0.40	
Chittenango Springs.	Resort.		
White Sulphur Spring.		0.08	
Cave Spring		0.22	
Crystal Springs	Resort.		0.58
Lebanon Thermal Spring	Used commercially and as a resort.	0.45	
Oak Orchard Acid Springs.			
Spring No. 1	Used commercially.	(6.51)	134.73
Spring No. 2	Do.		129.06
Oak Orchard Acid Water.	Do.	(1.92)	133.31
Richfield Springs.	Resort.		
Sulphur Spring		0.10	
Saratoga Springs	Do.		
Champion Spouting Spring	Used commercially.	0.46	
Congress Spring.	Do.	0.32	
Empire Spring	Do.	0.42	
Hathorn Spring.	Do.	0.13	
High Rock Spring.	Do.	1.22	
New Putnam Spring	Do.	0.22	
Union Spring.	Do.	0.32	
Vichy Spring.	Do.	0.48	

New Jersey.			
Schooley's Mountain Spring	Resort.	0.14	
Pennsylvania.			
Cresson Springs	Do.		
Magnesia Spring		0.01	
Gettysburg Lithia Spring	Used commercially.	0.02	
Guylyck and Gaylord's Spring	Do.	(0.17)	3.10
Blossburg Springs	Has considerable local reputation.	(1.97)	5.64
Maryland.			
Strontia Mineral Spring	Used commercially.	1.09	
Flint Stone Mineral Springs	Used locally		71.68
Virginia.			
Bath Alum Springs	Used commercially and as a resort.		
Spring No. 1		10.29	5.81
Spring No. 2		(9.00)	2.88
Spring No. 3		12.29	7.88
Clifton Springs	Resort.		
Spring No. 1		0.08	0.72
Spring No. 2			2.24
Blue Ridge Springs	Used commercially and as a resort.	0.14	
Farmville Lithia Springs			
Spring No. 2	Used commercially.	2.52	
Jordan Alum Springs	Resort.		
Chalybeate Spring		0.05	
Alum Spring		(7.61)	23.64
Spring No. 2		(3.36)	2.07
Spring No. 3		(2.06)	2.14
Spring No. 4		(24.32)	4.84

TABLE A.—CONTINUED.

Name of Springs and State in which they are Located.	Remarks.	Grains per Gallon.	
		Alumina.	Sulphuric Acid.
VIRGINIA.—Continued.			
Spring No. 5		(7.83)	7.90
Spring No. 6		(8.36)	5.32
Jordan's White Sulphur Springs	Used commercially and as a resort	0.01	
Kimberling Springs	Local resort.		
Red Sulphur Spring			0.17
Massanetta Mineral Springs	Used commercially and as a resort	0.16	
Pulaski Alum Springs	Do.	(6.48)	(1.98)
Rawley Springs	Do.		
Main Fountain		0.04	0.43
Orkney Springs	Resort.		
Bear Wallow Spring		0.01	0.36
Rockbridge Alum Springs	Used commercially and as a resort		
Chalybeate Spring		0.06	
Spring No. 1		14.76	18.79
Spring No. 2		17.91	15.22
Spring No. 3		43.95	2.04
Spring No. 4		24.09	5.51
Spring No. 5		(3.36)	2.07
Spring No. 6		(2.06)	2.14
Spring No. 7		(24.32)	4.84
Spring No. 8		(7.83)	7.90
Spring No. 9		(8.36)	5.32
Stribling or Augusta Springs	Used commercially and as a resort		
Alum Spring			9.09

No. 4 Alum Spring		(5.01)	5.05
No. 5 Alum Spring		(5.39)	9.82
No. 6 Alum Spring		(11.52)	6.54
Rock Enon Springs	Used commercially and as a resort	0.80	
Roanoke Red Sulphur Springs	Resort	0.01	
Shenandoah Alum Springs	Used commercially and as a resort	10.29	49.42
Variety Springs	Resort		
Alum Spring		(10.32)	1.37
Wallawhatoola Alum Springs		(21.63)	33.82
West Virginia.			
Capon Springs	Used commercially and as a resort		
Main Spring		0.02	
Beauty Spring		0.02	
Greenbrier White Sulphur Springs	Do.		
Sour Spring		37.42	112.80
Salt Sulphur Springs	Do.		
Iodine Spring		0.18	
North Carolina.			
Alum Spring of Onslow County	Resort		0.25
Cowhead Spring	Do.		1.23
Panacea Spring near Littleton	Used commercially and as a resort	0.32	0.43
Park's Alkaline Mineral Spring	Do.	3.50	
Georgia.			
Catoosa Springs	Resort and is beginning to be used commercially to some extent.		
No. 4 Chalybeate Spring		(0.20)	0.01
No. 3 Cosmetic Spring		(0.71)	0.01
No. 5 Magnesia Spring		(0.21)	0.01

TABLE A.—CONTINUED.

Name of Springs and State in which they are Located.	Remarks.	Grains per Gallon.	
		Alumina.	Sulphuric Acid.
GEORGIA.—Continued.			
No. 6 Congress Spring		(0.16)	0.01
No. 7 Alum Spring		(0.33)	0.01
No. 8 Black Sulphur Spring		(0.43)	0.12
No. 9 White Sulphur Spring		(0.74)	0.12‡
No. 10 Buffalo Spring		(0.71)	0.13‡
ALABAMA.			
Talladega Sulphur Spring	Resort.	1.45	315.85*
Roper Mineral Wells	Used commercially		
Johnston's Wells	Used locally	12.41†	39.45*
TENNESSEE.			
Austin's Springs	Resort.	2.00†	
Galbraith's Springs	Do.	0.04†	
Hurricane Springs	Sold to limited extent and a resort	0.29†	
Montvale Springs	Resort.	0.50†	
KENTUCKY.			
Kuttawa Mineral Springs	Has local reputation	70.08	
TEXAS.			
Kendall County Mineral Springs	Resort.		67.25
Mineral Well, Palo Pinto County	Used as a resort and to some extent commercially.	1.54†	
Sour Lake Mineral Springs	Used commercially and as a resort		

Name	Use		
Spring No. 7		(13.66)	16.67
Spring No. 9		(9.38)	6.18
Sour Springs, Caldwell County	Do.		7.26
Wootan Wells	Do.		
Well No. 4			86.41
OHIO.			
Green Mineral Spring	Used commercially and as a resort	0.98	
Ohio Magnetic Spring	Do.	0.12	
INDIANA.			
Greencastle Springs	Resort.		
Daggy Spring		0.16	
Dew Drop Spring		0.07	
West Saratoga Springs	Used commercially and as a resort		
Spring No. 1		0.18	7.26
Spring No. 2		0.36	3.20
Saint Ronan's Well	Resort.		245.80
ILLINOIS.			
Alcyone Mineral Springs	Used commercially.	1.65	
Glen Flora Springs	Used commercially and as a resort	0.15	
MICHIGAN.			
Butterworth's Magnetic Spring	Resort.	0.41	
Grand Haven Mineral Spring	Do.	0.30	
Mount Clemens Mineral Springs	Used commercially.		
Mineral Well		29.47	
Medea Spring		29.00	
Soolbad Spring		11.21	

† Aluminum Oxide. ‡ Sulphuric Acid (free). * Sulphuric Oxide.

TABLE A.—CONCLUDED.

Name of Springs and State in which they are Located.	Remarks.	Grains per Gallon.	
		Alumina.	Sulphuric Acid.
Wisconsin.			
Arctic Springs	Resort	0.15†	
Bethesda Spring, Waukesha	Used commercially and as a resort	0.12†	
Gihon Springs	Do.	7.59†	
Glenn Springs	Do.	0.05†	
Hacket's Spring	Unimproved, used locally	0.11†	
Horeb Mineral Spring	Used commercially and as a resort	0.23†	
Iodo-Magnesian Springs	Resort	0.06†	
Shealtiel Mineral Springs	Used commercially and as a resort	0.09†	
Sheboygan Mineral Spring, (artesian)	Used commercially	1.10†	
Sheridan Springs	Resort to small extent	0.05†	
Vesta Spring	Used commercially	0.13†	
White Rock Spring	Do.	0.62†	
Minnesota.			
Owatonna Mineral Springs	Resort		
Vichy Spring		0.10	
Name unknown		0.28	
Iowa.			
Cherokee Magnetic Mineral Springs	Used commercially and as a resort	0.29	
Chamberlain Mineral Springs	Local resort		25.02
Missouri.			
Bowsher Mineral Spring	Resort	0.65†	
Climax Springs	Do.		3.60

Landreth's Mineral Well	Do.	0.67†	
Sweet Springs, Brownville	Used commercially and as a resort		
Akesion Spring		0.17†	
Sweet Spring		0.09†	
KANSAS.			
Manhattan Artesian Mineral Wells	Used commercially and as a resort		
Well No. 1			61.36
Well No. 2			33.11
CALIFORNIA.			
Litton's Seltzer Springs	Used commercially and as a resort		2.36
Highland Springs	Resort		
Seltzer Spring		1.56	
Dutch Spring		0.11	
Magic Spring		0.17	
Tolenas Springs	Used commercially and as a local resort	0.96	
WASHINGTON.			
Medical Lake	Used commercially and as a resort	0.18†	

† Aluminum Oxide.

In some of the analyses given in the bulletin, the term "Alumina" is used and in others Aluminum Oxide. As these terms are identical in chemistry, I have placed the figures representing them in one column, headed "Alumina," but have indicated by daggers (†) those which are given in the analyses as Aluminum Oxide instead of "Alumina." It is quite probable, I think, that in many of the analyses given in the bulletin that the term "Sulphuric Acid" may have been used in a popular sense for Sulphur Trioxide or Sulphuric Oxide, which are identical. There are not any means of positively determining this from the bulletin, however, and I have, therefore, placed the figures representing them in one column headed "Sulphuric Acid," signifying by asterisks (*) and double daggers (‡) when the terms used in the analyses were not "Sulphuric Acid." The figures in parentheses in the column headed "Alumina," opposite to the figures representing "Sulphuric Acid," were estimated from figures representing Aluminum Sulphates, which I have assumed to be $Al_2 3 (SO_4)$. Other figures representing Aluminum Sulphate, given in the analyses in the bulletin, were not considered in compiling the table.

The results given in the bulletin relating to Hot or Warm Springs, are not included in the above table.

[As I have mentioned, in the main body of my report, about 0.5 of a grain of Basic Sulphate of Alumina was added to each gallon of Pawtuxet River Water during the experiments with the Experimental Morison Mechanical Filter, in order to purify it satisfactorily. It is shown on page 39 and in Professor Drown's report in the appendix, that 0.5 of a grain of a sample of Basic Sulphate of Alumina contained about *0.08 of a grain of Alumina* ($Al_2 O_3$) and about 0.18 of a grain of Sulphur Trioxide (SO_3). It is also shown in Professor Drown's report that a sample of Pawtuxet River Water contained per gallon, after filtration through paper, about 0.03 of a grain of Alumina ($Al_2 O_3$) and about 0.31 of a grain of Sulphur Trioxide (SO_3) (about 72 per cent. more than the sample of Basic Sulphate of Alumina contained), and that after adding 0.5 of a grain of Basic Sulphate of Alumina, per gallon, to the River Water and filtering through paper, the effluent water showed an increase in Alumina ($Al_2 O_3$) corresponding to the same quantity that was found in the River Water, and an increase of Sulphur Trioxide (SO_3) which corresponds to an amount about 33 per cent. less than was found in the River Water. I have quoted Professor Drown's report in making the above comparison, as Professor Appleton's analyses and our own tests indicate that there did not any of the Basic Sulphate of Alumina, that was added to the Applied Water, pass through the filter in its original state.]—E. B. W.

TABLE B.

The following table gives the quantity of "Alumina" and "Sulphuric Acid" contained in some Natural Waters in Massachusetts, compiled from the Report of the State Board of Health of Massachusetts for the year 1892. (The proportions of "Sulphuric Acid" in the table have been previously given in parts per 100,000 in the main portion of my report).

REMARKS.	GRAINS PER GALLON.	
	"Alumina." (Al_2O_3).	"Sulphuric Acid." (SO_3).
Normal Ground Waters.		
Mansfield, Well	0.01	0.08
Large Population on Drainage Area.		
Stoughton, Well	0.03	1.19
Everett, Spring	0.02	0.90
Malden, Tubular Wells	0.01	2.44
Improved Natural Filtration from Unpolluted Reservoir.		
Wayland, Reservoir	0.02	0.11
Wayland, Filter Gallery	0.03	0.08
Wells Containing Considerable Iron in Solution, as the Result of Organic Matter and Iron in the Ground.		
Westborough, Insane Hospital, Tubular Wells	0.01	0.30
Reading, Filter Gallery	0.03	3.48
Bradford, Well No. 7	0.01	0.41
Bradford, Well No. 12	0.01	0.22
Wells Near the Sea.		
Marblehead Water Company, Swampscott, Large Wells and Tubular Wells	0.01	1.74
Marblehead, Town Supply, Large Wells and Tubular Wells	0.01	1.79

TABLE C.

Showing the Number of Times, during 15 Winters, from 1880-81 to 1894-95 inclusive, that Periods occurred (Number of days without intermission), of 1 day and More, when the Daily Mean Temperature was 32° and Less, at Providence, R. I.

88	Periods of	1 day each.
59	Periods of	2 days each.
47	Periods of	3 days each.
21	Periods of	4 days each.
19	Periods of	5 days each.
13	Periods of	6 days each.
8	Periods of	7 days each.
5	Periods of	8 days each.
6	Periods of	9 days each.
2	Periods of	10 days each.
2	Periods of	11 days each.
4	Periods of	12 days each.
1	Period of	13 days.
1	Period of	14 days.
1	Period of	16 days.
1	Period of	17 days.
1	Period of	19 days.
1	Period of	22 days.

TABLE D.

Showing the Normal Mean January Temperature in Degrees F., of a number of European cities and at Providence, R. I. Also, the Normal Mean Temperature at Providence, R. I., for December and February.

NAMES OF CITIES.	December.	January.	February.
St. Petersburg		16°	
Warsaw		24°	
Königsberg		26°	
Providence, R. I	32.9°	27.5°	29.1
Zurich		29°	
Budapest		29°	
Posen		29°	
Frankfort-on-Oder		30°	
Berlin		31°	
Hamburg		31°	
Altona		31°	
Magdeburg		31°	
Bremen		32°	
London		38.5°	

TABLE E.

Showing the Number of Times and the Number of Days in each Period during each Winter from 1879-80 to 1886-87 and from 1888-89 to 1892-93, that the Hope Reservoir of the Providence Water Works was Frozen Over. Also, the Date that the Reservoir was Frozen Over the First Time during each Winter, and the Last Time that Ice was visible, and the Total Number of Days that the Daily Mean Temperature was 32° and Less.

Winter.	First Froze Over.	Last Ice Visible.	No. of Days in Each Period. 1st Time.	2d Time.	3d Time.	4th Time.	5th Time.	Total No. of Days.	Per cent. of Days Froze Over from First to Last.	Remarks.	No. of Days that Mean Temperature was 32° and Less.
1879–80	Dec. 23	Feb. 18	39	13	..	..	..	52	91	....	..
1880–81	Nov. 28	March 18	1	106	..	..	..	107	96	....	87
1881–82	Jan. 5	March 10	64	..	..	..	..	64	100	....	52
1882–83	Dec. 4	March 29	2	110	..	..	..	112	97	....	98
1883–84	Dec. 16	March 24	99	..	..	..	..	99	100	....	73
1884–85	Dec. 20	April 1	3	8	4	71	..	86	84	....	84
1885–86	Dec. 27	March 22	9	73	..	..	..	82	96	....	67
1886–87	Dec. 8	March 13	95	..	..	..	..	95	100	....	74
1887–88	......	....	..	..	..	..	..	..	..	Filling Reservoir.	72
1888–89	Dec. 15	March 4	2	17	5	1	28	53	67	....	51
1889–90	Jan. 17	March 13	3	13	4	9	..	29	53	....	39
1890–91	Dec. 8	Feb. 26	80	..	..	..	..	80	100	....	62
1891–92	Jan. 8	March 27	6	2	51	12	..	71	90	....	58
1892–93	Dec. 11	March 30	1	99	..	..	..	100	92	....	73
Averages.	Dec. 20	March 16	..	..	..	..	..	79	90	....	68

Cost of Filtration.

December 20, 1895.—The following estimates have been added to the appendix at the request of the Secretary of the State Board of Health, of Rhode Island :

Mechanical Filtration.—Estimates Nos. 1, 2, 3 and 4.

Estimates of the cost of four first-class Mechanical Filter Plants having an effective capacity each of 15,000,000 gallons per 24 hours, and the cost of operating the same when Basic Sulphate of Alumina (@ $0.02 per pound), is used, at the rate of $\frac{6}{10}$ of a grain per gallon of water filtered.

Each plant includes :—A Brick building, for housing filters and auxiliaries, having an iron roof and concrete floor, including smoke stack, flues, stairs, galleries and ladders,—A Wooden storage shed,—Cast-iron pipes and connections,—Gates and angle valves and wheel stands and wheels,—Centrifugal pumps, in duplicate,—Steam engines, in duplicate,—Boilers, in duplicate,—Boiler feed pumps and heater,—Electric lighting and signalling work,—Steam heating pipes, etc.,—Plumbing,—Chemical apparatus and connections,—Equipment of engine room,—Application of power, —Filters,—etc., etc.

Estimate No. 1.

Plant having *60 Steel Filters,* and based upon an average rate of filtration, when the entire number of filters are in service, of 100,000,000 gallons per Acre per 24 hours, as recommended in the report.

An actual rate of about 100,000,000 gallons would have required 58 filters, but as it was decided to arrange the filters in batteries of 3, there would have been 19 complete batteries and 1 odd filter. It was, therefore, thought best to add 2 more filters in order to have 20 complete batteries, and to be on the safe side in regard to a future increase of the consumption of water by the city. Assuming that 3 filters would be out of service the entire time on account of being washed, the remaing 57 would be obliged to filter 15,000,000 gallons per 24 hours, in addition to 735,000 gallons, the quantity required for washing the filters, and the average rate of filtration through the 57 filters while filtering 15,735,000, would be about 106,000,000 gallons per Acre per 24 hours. The reduction of the rate of filtration from 128,000,000, the average rate during the experiments, to 100,000,000 was not recommended for the purpose of obtaining purer water than could be procured at a rate of 128,000,000; but in order to have a sufficient reserve for washing the filters, and to enable the filters to be run a longer time between washings than 16h. 43m., the average length of time during the experiments (which could be brought about by reducing the rate of filtration), as it was thought that an increased length of time between washings might possibly be of advantage in the handling of a plant which would have a capacity many times larger than the experimental filter,—also, to assure beyond a possible doubt, if unforeseen difficulties should be encountered in the future, of the practical working of the filter plant and a positive delivery of 15,000,000 gallons of purified water per 24 hours, in addition to the quantity required for washing the filters, at a rate of filtration not to exceed under any circumstances, 128,000,000 gallons per Acre per 24 hours.

In the report I have given $5.69 as the cost per 1,000,000 gallons of operating a 15,000,000 Mechanical Filter Plant, including the cost of cleansing the filters twice a year with Caustic Soda. I have also mentioned in the report that I thought that the expense of cleansing the filters could be considerably reduced by the use of steam and other chemicals. Since 1893 I have made some careful investigations relative to cleansing large filters in practical operation, and I have found that the cost of cleansing can be very much reduced, from what it can be done for by using Caustic Soda, by the use of Soda Ash and Steam. In Estimates Nos. 1, 2, 3 and 4, I have, therefore, taken this reduction of cost into consideration, in addition to the use of unfiltered water as "rewash water" instead of filtered water, which reduces the cost of operating from $5.69 to $4.52.

Total Cost of filter plant $245,172,—interest on total cost, per annum @ 4 per cent., $9807,—annual deterioration of plant and repairs $7434,—cost considering the above figures $3.15 per 1,000,000 gallons filtered,—cost of operating, including cost of cleansing filters twice a year with "soda ash" and steam, $4.52 per 1,000,000 gallons.—*Total cost of filtration, $7.67 per 1,000,000 gallons.* (If 2 per cent. for a sinking fund was considered, instead of "deterioration," the total cost per 1,000,000 gallons would be $7.48).

Estimate No. 2.

Plant having *60 Seasoned Cypress Wood Filters*. The other conditions are the same as in Estimate No. 1.

Total Cost of filter plant $229,452,—interest on total cost, per annum @ 4 per cent., $9178,—annual deterioration of plant and repairs, $9132,—cost considering the above figures, $3.34 per 1,000,000 gallons filtered,—cost of operating, including cost of cleansing filters twice a year with "soda ash" and steam, $4.52 per 1,000,000 gallons.—*Total cost of filtration, $7.86 per 1,000,000 gallons.* (If 2 per cent. for a sinking fund was considered, instead of "deterioration," the total cost per 1,000,000 gallons would be $7.28).

Estimate No. 3.

Plant having *51 Steel Filters*, and based upon an average rate of filtration, through 48 of the 51 filters, of about 126,000,000 gallons per Acre per 24 hours (the average rate during the experiments being 128,000,000), while 48 of the filters are delivering an average quantity of 15,000,000 gallons per 24 hours, in addition to 735,000 gallons, the amount required for washing the filters.—It is assumed that the other 3 filters will always be out of service on account of being washed, etc. The average rate of filtration through the entire number of 51 filters, if they were all in service at the same time, while delivering an average quantity of 15,735,000 gallons per 24 hours, would be about 119,000,000 gallons per Acre per 24 hours.

Total Cost of filter plant $212,404,—interest on total cost, per annum @ 4 per cent., $8496,—annual deterioration of plant and repairs, $6445,—cost considering the above figures, $2.73 per 1,000,000 gallons filtered,—cost of operating, including cost of cleansing filters twice a year with "soda ash" and steam, $4.52 per 1,000,000 gallons.—*Total cost of filtration, $7.25 per 1,000,000 gallons.* (If 2 per cent. for a sinking fund was considered, instead of "deterioration," the total cost per 1,000,000 gallons would be $7.08).

Estimate No. 4.

Plant having *51 Seasoned Cypress Wood Filters*. The other conditions are the same as in Estimate No. 3.

Total Cost of filter plant $198,934,—interest on total cost, per annum @ 4 per cent., $7957,—annual deterioration of plant and repairs, $7888,—cost considering the above figures, $2.89 per 1,000,-000 gallons filtered,—cost of operating, including cost of cleansing filters twice a year with "soda ash" and steam, $4.52 per 1,000,000 gallons.—*Total cost of filtration, $7.41 per 1,000,000 gallons.* (If 2 per cent. for a sinking fund was considered, instead of "deterioration," the total cost per 1,000,000 gallons would be $6.91).

The cost of the filter plants considered in estimates Nos. 1, 2, 3 and 4, were based on actual figures given in a proposition for furnishing and constructing a large filter plant in 1893.

Natural Filtration.—Estimates Nos. 5, 6, 7 and 8.

On page 655 of the Report of the State Board of Health of Massachusetts, for the year 1894, may be found the following:

"*More Satisfactory Results from Covered Filters, in a Climate*" "*as exists at Lawrence.*"

"From experience with the out-door experimental Filters, No." "3 B and 8 A, and the Lawrence city filter it appears that the" "difficulty in scraping the surface during the winter months is" "so great that it is advisable to provide water filters with covers" "in this climate."

On April 18, 1895, the Water Board of Lawrence, Massachusetts, received a communication from the designer of the Two and One-half (2½) Acre Natural Filter-bed at Lawrence which first went into operation in September, 1893, recommending the covering of the filter-bed. Soon after, the Water Board requested the City Engineer of Lawrence to make plans to cover the bed. On June 28, plans were submitted to the Board, the estimated cost of which was $40,000. The covering which was designed for the filter-bed was to be of wood, with a roof to be covered with two-inch plank, with skylights, and sheathing inside, to make an air space to assist in preventing freezing.

The Total Cost charged to the construction of the Lawrence Filter, to January 1, 1894, was $69,531.74.—The Cost for Maintenance, labor and care of the Filter, during the year 1894, was $4614.50, and the Total Quantity pumped from the Filter during the year was 1,049,938,320 gallons. The Cost of Filtration, etc., per 1,000,000 gallons, during the year 1894, therefore, was $4.39.

From "The Filtration of Public Water-supplies, by Allen" "Hazen, Late Chemist in charge of the Lawrence Experiment" "Station of the Massachusetts State Board of Health," published in 1894, the following extracts relative to "Covers for Filters," have been taken:

"An addition to the Berlin filters, built in 1874, was covered" "with masonry vaulting. over which several feet of earth were" "placed, affording a complete protection against frost. The filters" "at Magdeburg built two years later were covered in the same way," "and since that time covered filters have been built at perhaps a" "dozen different places."

"Roofs have been used in Königsberg, Posen, and Budapest" "instead of the masonry vaulting. They are cheaper, but do not" "afford as good protection against frost, and even with great care" "some ice will form under them."

"To supply a maximum of 10,000,000 gallons daily, five filters" "each with an area of one acre will be ample. Any four of them" "can easily furnish this quantity while the fifth is out of use for" "cleaning or other cause. If the city is north of the line of nor-" "mal January temperature of 32°, vaulted filters will be required."

* * *

"Some estimates recently made by the author in connection" "with engineers examining the Boston Metropolitan Water-sup-" "ply indicate that filters fully up to the German standards, but" "with beds of a full acre each, and with vaulting substantially" "like that successfully used on the Newton covered reservoir," "can be built at present American prices for somewhat less than" "the cost given above, notwithstanding the higher price paid for" "American labor."

"Including the connection with the (existing) pumping-station" "we may estimate the cost of our five acres at $350,000, with a" "probability that with favorable local conditions the expenditure" "would be still less."

Estimate No. 5.

Filter-beds *Covered with Masonry Vaulting*, based upon Mr. Hazen's figures given above. 15,000,000÷10,000,000=1.5,— 1.5× $350,000=$525,000 as the Total Cost of the Filter-beds,—interest on total cost, per annum @ 4 per cent., $21,000,—deterioration and repairs, per annum, $3,500,—cost considering these figures, $4.47 per 1,000,000 gallons,—assumed cost of operating the filters, $4.39 per 1,000,000 gallons (the same as the cost at Lawrence for the year 1894).—*Total cost of filtration, $8.86 per 1,000,000 gallons.* (If 2 per cent. for a sinking fund was considered, instead of "deterioration," the total cost per 1,000,000 gallons would be $10.30).

Estimate No. 6.

Filter-beds *Covered with Wood,* based upon the figures given above relating to the two and one-half (2.5) acre filter-bed at Lawrence, Mass., and assuming the rate of filtration 2,000,000 gallons per Acre per 24 hours, should all of the beds be in service at the same time, and 2,500,000 when four-fifths (⅘) of the beds are in service, and one-fifth (⅕) out of use for cleaning or other cause. $69,532÷2.5=$27,813, cost of beds per acre,—$40,000÷2.5=$16,-000, cost of covering with wood per acre,—15,000,000÷2,000,000 =7.5,—cost of beds $27,813×7.5=$208,598,—cost of covering with wood, $16,000×7.5=$120,000,— Total Cost of filter-beds $208,598+$120,000=$328,598,—interest on total cost, per annum @ 4 per cent., $13,144,—deterioration and repairs per annum, $7391,—cost considering these figures, $3.75 per 1,000,000 gallons, —assumed cost of operating the filters, $4.39 per 1,000,000 gallons (the same as the cost at Lawrence for the year 1894).—*Total cost of filtration, $8.14 per 1,000,000 gallons.* (If 2 per cent. for a sinking fund was considered, instead of "deterioration," the total cost per 1,000,000 gallons would be $8.05).

Estimate No. 7.

Conditions the same as in Estimate No. 6, with the exception that the rate of filtration is assumed to be 2,500,000 gallons per Acre per 24 hours, should all of the beds be in service at the same time, and 3,125,000 when four-fifths (⅘) of the beds are in service, and one-fifth (⅕) out of service for cleaning or other cause. 15,-000,000÷2,500,000=6,—cost of beds $27,813×6=$166,878,—cost of covering with wood, $96,000,—Total Cost of filter-beds $166,878+ $96,000=$262,878,—interest on total cost, per annum @ 4 per cent., $10,515,—deterioration and repairs per annum, $5913,—cost considering these figures, $3.00 per 1,000,000 gallons,—assumed cost of operating the filters, $4.39 per 1,000,000 gallons (the same as the cost at Lawrence for the year 1894).—*Total cost of filtration, $7.30 per 1,000,000 gallons.* (If 2 per cent. for a sinking fund was considered, instead of "deterioration,", the total cost per 1,000,000 gallons would be $7.32).

ESTIMATE NO. 8.

The following estimate is based upon my own figures:—Ten (10) filter-beds, not covered, having an area of 0.94 of an Acre each, to filter at a rate not to exceed 2,000,000 gallons per Acre per 24 hours, and to have a depth of water over the top of the beds of about 4 feet. Eight (8) of these beds to filter at the rate of 2,000,000, and to have a capacity of 15,000,000 gallons per 24 hours. The other two (2) filter-beds are to be held in reserve for cleaning or other cause.

Total Cost of the filter-beds $291,220, assuming that a suitable quality of sand, which would not need to be washed nor require much screening, could be obtained for the filtering medium in the immediate vicinity of the location of the filter-beds,—interest on total cost, per annum @ 4 per cent., $11,649,—deterioration and repairs per annum, $1,941,—cost considering these figures, $2.48 per 1,000,000 gallons,—assumed cost of operating the plant, $4.39 per 1,000,000 gallons (the same as the cost at Lawrence for the year 1894).—*Total cost of filtration, $6.87 per 1,000,000 gallons.* (If 2 per cent. for a sinking fund was considered, instead of "deterioration," the total cost per 1,000,000 gallons would be $7.67).

SUMMARY OF THE ABOVE ESTIMATES.

The figures in parentheses include 2 per cent. for a sinking fund instead of "deterioration."

MECHANICAL FILTRATION.	Total Cost of Plant.	Cost per 1,000,000 Gallons Filtered.	
Estimate No. 1.—Including 60 "Steel" Filters	$245,172	$7.67	($7.48)
" No. 2.—Including 60 "Cypress" Filters	$229,452	$7.86	($7.28)
" No. 3.—Including 51 "Steel" Filters	$212,404	$7.25	($7.08)
" No. 4.—Including 51 "Cypress" Filters	$198,934	$7.41	($6.91)

Natural Filtration.	Total Cost of Plant.	Cost per 1,000,000 Gallons Filtered.	
Estimate No. 5.—Filter-beds covered with Masonry Vaulting, Rate of Filtration 2,000,000 and 2,500,000...........	$525,000	$8.86	($10.30)
" No. 6.—Filter-beds covered with Wood, Rate of Filtration 2,000,000 and 2,500,000.......	$328,598	$8.14	($8.05)
" No. 7.—Filter-beds covered with Wood, Rate of Filtration 2,500,000 and 3,125,000.......	$262,878	$7.39	($7.32)
" No. 8.—Filter-beds Not Covered, Rate of Filtration 2,000,000.......	$291,220	$6.87	($7.67)

In the estimates of each of the Mechanical Filter Plants, I have assumed that the ground upon which they would be erected was graded, and of a character suitable for the foundation of the building, etc., to rest upon, and in the estimates of the Natural Filter-beds I have assumed that the excavation would be suitable for the construction of the embankments and that not any more than the ordinary difficulties of construction would be encountered. In any of the estimates I have not considered the cost of settling basins, if such should be needed, nor pipes nor conduits leading to and from the filters, etc., etc. Also, I have not considered the cost of land in any of the estimates, which would be of considerable importance in some localities and probably exert more or less influence in the selection of a filter-plant, as, for instance, the Mechanical Filter Plant considered in Estimate No. 1 could be enclosed within an area of less than 1 Acre, while the Natural Filter-beds considered in Estimate No. 8, would require in all probability at least an area of 16 Acres for their site.—E. B. W.

www.ingramcontent.com/pod-product-compliance
Lightning Source LLC
Chambersburg PA
CBHW020847160426
43192CB00007B/814

9783337139087